APPROACHES FOR EVALUATING THE NRC RESIDENT RESEARCH ASSOCIATESHIP PROGRAM AT NIST

Board on Higher Education and Workforce
Policy and Global Affairs Division

John Sislin, Editor

NATIONAL RESEARCH COUNCIL
OF THE NATIONAL ACADEMIES

THE NATIONAL ACADEMIES PRESS
Washington, D.C.
www.nap.edu

THE NATIONAL ACADEMIES PRESS 500 Fifth Street, N.W. Washington, D.C. 20001

This project was supported by the National Institute of Standards and Technology, Grant No. SB1341-04-C-0001. Any opinions, findings, conclusions, or recommendations expressed in this publication are those of the author(s) and do not necessarily reflect the views of the organizations or agencies that provided support for the project.

International Standard Book Number-13: 978-0-309-11218-5
International Standard Book Number-10: 0-309-11218-4

Additional copies of this report are available from the National Academies Press, 500 Fifth Street, N.W., Lockbox 285, Washington, D.C. 20055; (800) 624-6242 or (202) 334-3313 (in the Washington metropolitan area); Internet, http://www.nap.edu.

Suggested citation: National Research Council. 2007. Approaches for Evaluating the NRC Resident Research Associateship Program at NIST. Board of Higher Education and Workforce. John Sislin, ed. Washington, D.C.: The National Academies Press.

Printed in the United States of America

THE NATIONAL ACADEMIES

Advisers to the Nation on Science, Engineering, and Medicine

The **National Academy of Sciences** is a private, nonprofit, self-perpetuating society of distinguished scholars engaged in scientific and engineering research, dedicated to the furtherance of science and technology and to their use for the general welfare. Upon the authority of the charter granted to it by the Congress in 1863, the Academy has a mandate that requires it to advise the federal government on scientific and technical matters. Dr. Ralph J. Cicerone is president of the National Academy of Sciences.

The **National Academy of Engineering** was established in 1964, under the charter of the National Academy of Sciences, as a parallel organization of outstanding engineers. It is autonomous in its administration and in the selection of its members, sharing with the National Academy of Sciences the responsibility for advising the federal government. The National Academy of Engineering also sponsors engineering programs aimed at meeting national needs, encourages education and research, and recognizes the superior achievements of engineers. Dr. Charles M. Vest is president of the National Academy of Engineering.

The **Institute of Medicine** was established in 1970 by the National Academy of Sciences to secure the services of eminent members of appropriate professions in the examination of policy matters pertaining to the health of the public. The Institute acts under the responsibility given to the National Academy of Sciences by its congressional charter to be an adviser to the federal government and, upon its own initiative, to identify issues of medical care, research, and education. Dr. Harvey V. Fineberg is president of the Institute of Medicine.

The **National Research Council** was organized by the National Academy of Sciences in 1916 to associate the broad community of science and technology with the Academy's purposes of furthering knowledge and advising the federal government. Functioning in accordance with general policies determined by the Academy, the Council has become the principal operating agency of both the National Academy of Sciences and the National Academy of Engineering in providing services to the government, the public, and the scientific and engineering communities. The Council is administered jointly by both Academies and the Institute of Medicine. Dr. Ralph J. Cicerone and Dr. Charles M. Vest are chair and vice chair, respectively, of the National Research Council.

www.national-academies.org

Committee on Approaches for the Evaluation of the NIST/NRC Postdoctoral Research Associateships Program

Isaac Sanchez, Ph.D., (Chair), William J. Murray Endowed Chair in Engineering, Department of Chemical Engineering, University of Texas

Burt Barnow, Ph.D., Associate Director, Institute for Policy Studies, Johns Hopkins University

Kathryn Newcomer, Ph.D., Director of the Ph.D. in Public Policy and Administration program and Associate Director of the School of Public Policy and Public Administration, George Washington University

Georgine Pion, Ph.D., Research Associate Professor, Department of Psychology and Human Development, Peabody College, Vanderbilt University

Study Staff
Peter Henderson, Board Director
John Sislin, Study Director
Jim Voytuk, Senior Program Officer
Kara Murphy, Program Assistant
Rae Allen, Administrative Assistant

Board on Higher Education and Workforce

Ronald Ehrenberg, Ph.D., (Chair), Irving M. Ives Professor, Industrial and Labor Relations and Economics, Cornell University

Burt Barnow, Ph.D., Associate Director, Institute for Policy Studies, Johns Hopkins University

Donald L. Bitzer, Ph.D., Distinguished University Research Professor, Computer Science Department, North Carolina State University

Carlos G. Gutierrez, Ph.D., Professor of Chemistry, Department of Chemistry and Biochemistry, California State University

Donald Johnson, Ph.D., Vice President (retired), Product and Process Technology, Grain Processing Corporation

Claudia Mitchell-Kernan, Ph.D., Vice Chancellor of Graduate Studies and Dean, Graduate Division, University of California, Los Angeles

Michael Nettles, Ph.D., Edmund W. Gordon Chair for Policy Evaluation and Research Educational Testing Service

Debra Stewart, Ph.D., President, The Council of Graduate Schools

Tadataka Yamada, M.D., Chairman, Research and Development, GlaxoSmithKline

Staff
Peter Henderson, Board Director
Jim Voytuk, Senior Program Officer
John Sislin, Program Officer
Jim McKinney, Financial Associate
Kara Murphy, Program Assistant
Rae Allen, Administrative Assistant

FIGURES

BOXES

Executive Summary

The NRC Resident Research Associateship Program (RAP) of the National Institute of Standards and Technology (NIST), hereafter "NIST/NRC RAP," was started in 1954. The program provides two-year temporary appointments at NIST for outstanding scientists and engineers chosen through a national competition. These appointments are designed to provide an opportunity for some of the nation's best scientists, mathematicians, and engineers to engage in state-of-the-art research in association with senior research specialists of NIST's staff and to make use of the well-regarded and often unique research facilities at NIST. The RAP is perceived to provide multiple benefits to the postdoctoral research associates, to NIST, and to the scientific and engineering community at large. NIST approached The National Academies with a desire to see what sort of evaluation could be undertaken given available data. In addition, NIST was interested in recommendations for future data collection where data were found to be currently lacking, and for more in-depth evaluation strategies that could be done in the future. Based on this request from NIST, an ad hoc study committee—the Committee on Approaches for the Evaluation of the NIST/NRC Postdoctoral Research Associateships Program—was appointed by the National Research Council to conduct a study.

The committee's specific charge is presented below:

> "The Academic Competitiveness Council, in furtherance of the Administration's American Competitiveness Initiative, seeks to ensure that the nation invests wisely and effectively in educational programs to meet its science and technology goals. The ACC, therefore, requires evaluations of important STEM[1] education programs, including the NIST/NRC Postdoctoral Research Associateship Program. An ad hoc committee under the auspices of the Board on Higher Education and Workforce will describe the pool of applicants for and recipients of the NIST/NRC Postdoctoral Research Associateship Program and carry out a descriptive analysis of career outcomes of NIST postdoctoral scholars based on available information. As possible given available data, the committee will also describe how the program addresses agency goals. The committee will also outline an approach to evaluating the program relative to national S&E goals, NIST agency goals, and the value of the program to participants, which could be undertaken in a future study."

To meet its charge, the committee focused on three objectives: (1) to describe characteristics of NIST applicants compared to the general pool of new science and engineering doctorates; (2) to describe the experiences of Research Associates at NIST compared to other Research Associates in other programs; and (3) to offer suggestions for conducting a more in-depth assessment of the careers of Research Associates at NIST, with a particular focus on quantifying the benefits of the appointment to the recipients as well as to NIST—during and after the appointment period. The committee was guided by two principal questions:

(1) Is NIST attracting the "best and the brightest" to the Research Associateship Program?

---

[1] STEM is science, technology, engineering, and mathematics.

(2) What is the impact of the Program on the Research Associates, NIST, and relevant scientific fields in general?[2]

In addition, the committee determined to offer recommendations where appropriate regarding data collection on applicants to the Program; on the experiences of current and former NIST Research Associates; and on the views of Research Associates and NIST employees towards the value of the Program to Research Associates, to NIST, and to science and engineering broadly.

Multiple sources of information were identified and used in preparing this report. The primary source of information was data collected by the NRC's Fellowships Office, including: application data, data from final reports prepared by the Research Associates, and a directory of past Research Associates. In addition, the committee collected original data from three expert panels. In general, however, the data were inadequate to draw definitive conclusions.

This report is divided into five chapters. Chapter One describes the program and the approach and scope of the study. Chapter Two examines applicants to the NIST/NRC RAP and compares them to applicants to other Research Associateship Programs. It also examines applications and awards disaggregated across several dimensions, such as gender or doctoral-granting institution. Chapter Three examines the experiences of NIST/NRC Research Associates and Research Associates at other federal agencies, as well as Research Associates' views on the value of the program they participated in. Chapter Four examines the careers of former Research Associates. Chapter Five presents an overall summary of preliminary results and recommendations.

PRELIMINARY RESULTS BASED ON AVAILABLE DATA

1. **Outreach efforts produce more qualified applicants than NIST has slots to fill for Research Associates; and the pool of applicants includes many from top research institutions and is increasingly diverse.** Overall, 22 percent of applicants to NIST were awarded an appointment—a lower awards ratio than for other RAPs overall. The award ratios for NIST and other RAP applicants vary by gender, race, and field. Across all fields, the proportion of women and underrepresented minorities in the applicant pool and as awardees has grown over time, however less so than the proportion among Ph.D.s and those intending to be postdoctorates. Personal communication is the primary way that NIST/NRC Research Associates heard about the program.
2. **NIST/NRC Research Associates appear to be about as productive as Research Associates in other Programs.** On average, NIST/NRC Research Associates publish about two articles, give about four presentations, but rarely receive a patent or award during their appointments. They are more likely to give a domestic presentation or win an award, less likely to publish journal articles, and as likely to receive a patent or give an international presentation. Productivity data, though, are derived from a survey with a low response rate and possible nonresponse bias.
3. **Research Associates are quite satisfied with the program.** For those Research Associates who provided information on their final reports, on a scale of 1 to 10, with 10 being excellent, NIST/NRC Research Associates rated short-term and long-term value of the program; lab, advisor, administrative (NIST and NRC) support between 7.7 and 8.5.

[2] Given the limited data available and the charge to the committee, the committee could not provide a full assessment of the impact of the program in this report.

In half the categories NIST/NRC Research Associates and Research Associates in other programs reported statistically similar levels of satisfaction. In the other half, other Research Associates reported higher levels of satisfaction. Satisfaction data, though, are derived from a survey with a low response rate and possible nonresponse bias.

4. **Research Associates contribute to the pool of qualified applicants to permanent positions at NIST.** Among Research Associates completing their awards, 45 percent of those at NIST reported that their immediate post-tenure position was as a permanent, temporary, or contract employee—a higher percentage than Research Associates at other federal agencies. The employment of nonrespondents remains unknown. A survey of former Research Associates found that a higher percentage of former NIST/NRC Research Associates stayed at NIST than Research Associates in other programs stayed at their host agency (37.6 to 28.1 percent), among those former Research Associates who could be located.

CONCLUSION

1. **Currently available data do not allow for a full program evaluation.** Currently, the most thorough data are collected on applicants. Little data are collected on Research Associates' experiences; research advisors' evaluation of Research Associates; career outcomes of former Research Associates; and the value of the program to NIST or to the broader scientific and engineering community.

RECOMMENDATIONS

1. **NIST should conduct a more thorough evaluation of the NIST/NRC Research Associateship Program.**
    a. As a first step, NIST and the NRC should review specific goals of the program.
    b. The evaluation should include the following components: an assessment of outreach to potential applicants; an assessment of individuals who decline to accept a Research Associate position; an assessment of the benefits of the program on the Research Associates after they complete their appointments; an assessment of the benefits to NIST of hosting Research Associates; and an assessment of benefits of the program to the broader scientific and engineering community.
2. **NIST should conduct an evaluation of outreach efforts.**
    a. Additional analysis could be undertaken on how applicants hear about the program (e.g., focusing on the "Other" category).
    b. Additional data could be collected from NIST personnel and former or current NIST Research Associates. Such data could be used to answer such questions as how NIST personnel and Research Associates interact with potential applicants and which mechanisms seem to work best.
    c. A second step to facilitate an evaluation of outreach efforts is to identify metrics for quantifying value obtained from different outreach strategies.
    d. Examine individual outreach strategies for return on investment.

   e. Finally, consider whether there might be other outreach strategies that are being underused currently, and which might have potential value, such as direct mail to deans, department heads and other university administrators.

3. **NIST should conduct an evaluation of individuals who decline offers of research associateships.** This could be done as a telephone interview or via a survey. Two basic questions should be asked of those who are awarded but decline: (1) why are you declining, and (2) what are you planning to do instead?
4. **The NRC should amend the application form.** The list of fields should be reduced, in particular by collapsing very similar labels and by removing labels that are for multiple fields.
5. **The NRC should update the DataRAP database to replace organizational names (e.g., institutes or labs) that no longer exist at NIST with current equivalents.**
6. **NIST should conduct a more thorough assessment of Research Associates' experiences during the postdoctoral appointment, their satisfaction with and views on the benefits of the program, and NIST staff's satisfaction with and views on the benefits of the program.**
   a. To assist in this, the NRC should redesign the final report and the Research Advisor's evaluation form to maximize the collection of data from these instruments.
   b. The final report and the Research Advisor's evaluation should be made mandatory.
   c. Some elements of the current data collected could be subjected to further analysis.
      i. NIST may wish to conduct further analysis on peer-reviewed journals.
      ii. NIST may wish to conduct an impact analysis of Research Associates' productivity.
      iii. NIST may wish to conduct a more thorough review of their support of Research Associates, asking how familiar they are with NIST administrative offices, how often they turn to those offices for help, and for what reasons.
   d. NIST could also conduct a social network analysis of the collaboration of the Research Associates (or of NIST employees) to see how the Research Associateship Program facilitates new or wider collaboration among scientists and engineers.
   e. When data allow, NIST could consider disaggregating productivity and satisfaction measures for Research Associates by lab, gender, and race/ethnicity.
7. **NIST should conduct a broad evaluation of the careers of former Research Associates to evaluate the impact of the Program on Research Associates' careers, NIST, and the broader science and engineering community**. The best approach for doing this is a survey, which would compare the career outcomes of NIST/NRC Research Associates to similar postdocs. The survey would be directed towards these former Research Associates and a suitable control group. Ideally, two possible comparisons could be made. First, one could construct a peer group. This would consist of a matched or stratified sample of individuals who had postdocs similar to the one at NIST for the comparison group. Although not ideal, one solution would be to take a stratified sample of former Research Associates from the Fellowships Office's Directory. This is a census of former Research Associates; but as noted earlier in the report, many of these

individuals could not be found or failed to respond to an earlier survey designed to collect information on their current employment. A second comparison group would consist of similar doctorates. A roster could be assembled by tapping the group of applicants to RAPs, who did not receive an award. These individuals will likely exhibit a diversity of career paths, including some who took postdocs (in academia or industry) and others who went straight into employment.

1

Overview

In *Rising Above the Gathering Storm*, the National Academies posited that "The prosperity the United States enjoys today is due in no small part to investments the nation has made in research and development at universities, corporations, and national laboratories over the last 50 years. Recently, however, corporate, government, and national scientific and technical leaders have expressed concern that pressures on the science and technology enterprise could seriously erode this past success and jeopardize future U.S. prosperity"(NAS/NAE/IOM, 2007).

To address these challenges, the *Gathering Storm* report recommended action in four areas: K–12 education, higher education, science and engineering research, and economic and technology policy. Many of the report's recommendations were later echoed in the American Competitiveness Initiative (ACI), introduced by President Bush in his State of the Union Address on January 31, 2006. The ACI was warmly received by legislators from both parties on Capitol Hill, generating bills designed to implement its key provisions.

While considering these proposals, Congress took the immediate step of requiring an examination of existing federal education programs in science, technology, engineering, and mathematics (STEM) fields. As noted in a statement by Secretary Spellings on the new Academic Competitiveness Council (ACC) and its mission, "The Deficit Reduction Act, signed into law by the President on February 8, 2006, included an Academic Competitiveness Council chaired by the Secretary of Education, and consisting of members of the Federal Government whose agencies have education programs in science, technology, engineering, and mathematics. Its mission under law is to evaluate the effectiveness of each program, identifying areas of overlap and recommending ways to efficiently integrate and coordinate in the future." The ACC's efforts included assessing the success of these programs, identifying areas for improvement for current and future programs, and laying the groundwork for sustained collaboration among the programs (DOE, 2007).

Postdoctoral programs[3]—of which there are several in the sciences and engineering (S&E)—are of critical importance in providing additional research training, knowledge, and opportunities to bridge higher education with the start of a career in S&E. While many Ph.D.s do go straight into full-time employment, a large number of doctorates view a few years as a postdoc to be a valuable opportunity to build skills and a reputation and a chance to experiment in the labor market.

FEDERAL PROGRAMS TO SUPPORT POSTDOCS IN S&E

According to the GAO (2005:3), there were 207 "education programs funded in fiscal year 2004 that were designed to increase the numbers of students and graduates pursuing STEM[4] degrees and occupations or improve educational programs in STEM fields…." Included in these

---

[3] A postdoctoral position or "postdoc" is a temporary position awarded in academia, industry, non-profits, or government primarily for gaining additional education and training in research. Individuals with postdoctoral appointments are called postdocs.

[4] STEM is science, technology, engineering, and mathematics.

programs are several programs that provide and emphasize postdoctoral opportunities, such as: the National Aeronautics and Space Administration's NASA Postdoctoral Program; the National Institute of Health's Ruth L. Kirschstein National Research Service Awards (NRSA) for Individual Postdoctoral Fellows (F32), Career Transition Award (K22), Pathway to Independence (PI) Award (K99), Postdoctoral Intramural Research Training Award (IRTA), Cancer Research Training Award (CRTA), Postdoctoral Visiting Fellowship (VF), National Institute of General Medical Sciences Pharmacology Research Associate (PRAT) Program, and Women's Health Postdoctoral Fellowship; the National Science Foundation's Postdoctoral Research Fellowships; and the U.S. Department of Agriculture's Agricultural Research Service (ARS) Postdoctoral Research Associate Program.[5]

The National Research Council Fellowships Office of The National Academies manages several postdoctoral research programs on behalf of federal agencies. Table 1-1 lists several programs administered by National Research Council Fellowships Office.

---

[5] While these programs focus on individual postdocs, other programs target academic institutions with funding to support groups of postdocs.

TABLE 1-1 Selected National Research Council Research Associateship Programs

Agency	Research Associateship Program
Department of Defense	Air Force Research Laboratory (AFRL)
Department of Defense	Armed Forces Radiobiology Research Institute (AFRRI)
Food and Drug Administration/HHS	Center for Devices and Radiological Health (CDRH)
Defense Threat Reduction Agency/DOD	Chemical and Biological Defense (CBD)
Federal Aviation Administration/DOT	FAA/Civil Aerospace Medical Institute (CAMI)
Federal Highway Administration/DOT	Federal Highway Administration/Turner-Fairbank Highway Research Center (FHWA)
US Army Corps of Engineers/DOD	Institute for Water Resources (IWR)
Department of Energy	National Energy Technology Laboratory (NETL)
Department of Energy	National Energy Technology Laboratory Methane Hydrates Fellowship Program (NETL/MHFP)
Health & Human Services	National Institute for Occupational Safety and Health (NIOSH)
Department of Commerce	National Institute of Standards and Technology (NIST)
Health & Human Services	National Institutes of Health (NIH)
HHS and DOC	National Institutes of Health (NIH)/National Institute of Standards and Technology (NIH/NIST)
Department of Commerce	National Oceanic and Atmospheric Administration (NOAA)
Department of Defense	Naval Medical Research Center/Naval Health Research Center (NMRC/NHRC)
Department of Defense	Naval Postgraduate School (NPS)
Department of Defense	Naval Research Laboratory (NRL)
Department of Defense	Navy Marine Mammal Program (MMP)
Pacific Disaster Center	Pacific Disaster Center (PDC)
Department of Defense	US Army Aviation and Missile Command (AMCOM)
Department of Defense	US Army Edgewood Chemical Biological Center (ECBC)
Department of Defense	US Army Medical Research and Materiel Command (AMRMC)
Department of Defense	US Army Natick Soldier Research, Development and Engineering Center-US Army Research, Development and Engineering Command (NSRDEC)
Department of Defense	US Army Research Laboratory (ARL)
Department of Defense	US Army Research Office (ARO)
Department of Defense	US Army Research, Development & Engineering Command, Night Vision & Electronic Sensors Directorate (NVESD)
Department of Defense	US Army Research, Development, and Engineering Command/Armament Research Development and Engineering Center (RDECOM/ARDEC)
Environmental Protection Agency	US Environmental Protection Agency (EPA)
Department of the Interior	US Geological Survey (USGS)
US Marine Mammal Commission	US Marine Mammal Commission (MMC)
Department of Defense	US Military Academy/US Army Research Laboratory (USMA/ARL)

Note: Excluded were programs directed solely toward pre-doctoral students or faculty.
Source: NRC Associateships Research Opportunity Directory.

One concern, as noted by both the GAO (2005) and the Academic Competitiveness Council (ACC) (2007), was that not all programs have been evaluated, and even in cases of completed or ongoing evaluations, not much is known about the beneficial impact of individual programs.[6] Such evaluation is important and can illuminate a number of impacts, including:

---

[6] However, there have been several program evaluations of federal S&E programs and the authors of those evaluations might disagree that the evaluations have not produced information about the impact of those programs.

- The program's success in fostering diversity across gender, race/ethnicity, and/or socioeconomic status;
- The effectiveness of an agency's information dissemination strategies;
- Benefits to the recipients, the agency, and the nation as measured by the quality of the scholars' experiences in the program; and
- Benefits to recipients, the agency, and the nation as measured by the careers of postdocs.

It was in this spirit that the National Institute of Standards and Technology (NIST) approached The National Academies to undertake a study of the NRC Resident Research Associateship Program at NIST (hereafter "NIST/NRC RAP").

THE NRC RESIDENT RESEARCH ASSOCIATESHIP PROGRAM AT NIST

As noted above, the Fellowships Office of the National Research Council (NRC) of the National Academies administers a series of postdoctoral programs at several federal agencies. Collectively, these programs are known as the NRC Resident Research Associateship Programs. The Programs were established in 1954, modeled after the Rockefeller Foundation Fellowship program administered by the NRC from 1919 to 1955. NIST was the first sponsor: "The initial sponsorship of the RAP was through the National Bureau of Standards (now NIST) in 1954. NIST was joined by the Naval Research Laboratory in 1955 and other sponsors soon followed. The Research Associateship programs have continued to expand to the present day where over 30 federal agencies participate."[7] The NRC has made more than 1,000 awards in the NRC Resident Research Associateship Program at NIST. (Overall, the NRC has made 11,000 awards to postdoctoral and senior scientists and engineers to conduct research in federal laboratories.)

A NRC Resident Research Associate at NIST is a researcher and a term (temporary) employee of NIST. Associateships are analogous to fellowships or similar temporary employment programs at the postdoctoral level in universities and other organizations. Awardees are offered two-year term civil service appointments. During their tenure, Associates conduct research in one of six areas of interest to NIST: chemistry, computer science, engineering, materials science, mathematics, and physics. Associates devote their full-time effort to the research program proposed in their applications and are in residence at the sponsoring laboratory at NIST during the entire period of the Associateship.[8]

According to the NRC, the objectives of the programs are "(1) to provide postdoctoral scientists and engineers of unusual promise and ability opportunities for research on problems, largely of their own choice that are compatible with the interests of the sponsoring laboratories and (2) to contribute thereby to the overall efforts of the federal laboratories. For recent doctoral graduates, the programs provide an opportunity for concentrated research in association with

---

See for example: OSIRE (2003), McCullough and Thurgood (2004), Kimsey in NRC (2006a), Mantovani et al. (2006), Michie et al. (2007), and Finkelstein and Libarkin (undated).

[7] This information is adapted from The National Academies' RAP website, "Mission," available at: http://www7.nationalacademies.org/rap/RAP_Mission_Statement.html. [Accessed June 24, 2007.]

[8] However, during the expert panels held with NIST/NRC Research Associates (see description at the end of this chapter), it was noted that some Research Associates work on research projects other than what they proposed in their application packages. This fact is independent of whether such change is a positive or negative for the Research Associate. It is not clear how often this happens.

selected members of the permanent professional laboratory staff, often as a climax to formal career preparation. Participating laboratories receive a stimulus to their programs by the presence of bright, highly motivated, recent doctoral graduates with records of research productivity. New ideas, techniques, and approaches to problems contribute to the overall research climate of the laboratories. Indirectly, associateships also make available to the broader scientific and engineering communities the excellent and often unique research facilities that exist in federal laboratories."[9] Associates are encouraged to publish their research in refereed science and engineering journals, and many also present papers at U.S. and international conferences.

The objectives for each agency participating in the Research Associateship Programs may go beyond the NRC's goals, emphasize different aspects, or vary from the objectives of other agencies. The objectives of the NIST/NRC RAP, as stated by NIST, are: "The postdoctoral program brings research scientists and engineers of unusual promise and ability to perform advanced research related to the NIST mission, introduces the latest university research results and techniques to NIST scientific programs , strengthens mutual communication with university researchers, shares NIST unique research facilities with the U.S. scientific and engineering communities, and provides a valuable mechanism for the transfer of research results from NIST to the scientific and engineering communities."[10]

Additional objectives for NIST might include attracting a sufficient number of qualified candidates, increasing the pool of potential employees, increasing the breadth or depth of research capability at NIST, or increasing the productivity of NIST,[11] NIST may also be interested in whether employees who were research associates have better outcomes than employees who were not in the program and the effect of the program on NIST's mission and reputation over the long-term.

Finally, there are objectives for those who apply to the NIST/NRC RAP, including increased collaboration or networking, productivity, training, skills, knowledge, ability to work independently, or career choices.[12] Many former NIST/NRC RAs have gone on to distinguished careers: "Ten high level managers who now work at NIST were past postdocs.

---

[9] NRC, "NRC Associateships Research Opportunities Directory." Available at: http://nrc58.nas.edu/pgasurvey/data/aobooks/rapbooks.asp?mode=frntmtr&progctr=50&seq=20.

[10] Office of Academic Affairs, "Office of Academic Affairs." Available at: http://www.nist.gov/oiaa/acdmaffr.htm.

[11] During the expert panel with NIST advisors and managers, participants commented that the research associates were a way for NIST to cover a wider range of expertise. The postdoc program brings in staff to do research. It also brings in new ideas. Research Associates at NIST do some of the innovative research. The NRC notes that participating laboratories receive a stimulus to their programs by the presence of bright, highly motivated, recent doctoral graduates with records of research productivity. New ideas, techniques, and approaches to problems contribute to the overall research climate of the laboratories. Indirectly,associateships also make available to the broader scientific and engineering communities the excellent and often unique research facilities that exist in federal laboratories.

[12] During the expert panel with former research associates, participants suggested benefits to the program included increased collaboration (including the range of people you work with) and networking. The postdoctoral position was seen as a helpful step to the next career choice. Participants mentioned that the program acclimated postdocs to working in the government. In some areas within NIST, the postdoctoral position seemed to the participants to be the primary way to get a full-time job there—former postdocs assumed this because in some areas almost everyone was a former research associate. During the expert panel with NIST advisors and managers, participants commented that the postdoc is a way for people to explore working at NIST—doctorates might not want to come in as a full-time regular appointment, but would be interested in trying the place out for 2 years. NIST advisors and managers also noted that the Program is a good way for NIST to "try out" postdocs who might later become employees.

Three members of the National Academy of Engineering and the National Academy of Sciences were past NIST postdocs. Four of the present 27 NIST Fellows are past postdocs (Curry, 2004)."[13]

Current application requirements include:

- Research Associateships at NIST are open only to citizens of the United States; Permanent residency status is not sufficient;
- Research Associateships at NIST are awarded only to persons who have held the doctorate less than five years at the time of application;
- Awardees must hold the Ph.D., Sc.D., or other earned research doctoral degree recognized in U.S. academic circles as equivalent to the Ph.D. or must present acceptable evidence of having completed all the formal academic requirements for one of these degrees before tenure may begin;
- Applicants must have demonstrated superior ability for creative research; and
- A primary objective of the associateship programs is to provide a mechanism for new ideas and sources of stimulation to be brought to the sponsoring laboratory. Thus, persons with recent prior affiliation with a specific laboratory may not be eligible to apply for an associateship there.[14]

Reapplication is possible. Persons who have previously held an associateship may apply for another award only if a period of at least two years will have elapsed between termination of the first award and the proposed tenure of a second. Persons who have previously applied for an associateship, but who were not recommended for an award by the panels, may reapply after one year. Candidates who were recommended for an award by the panels, but who were not offered an award because of funding or other limitations, may reapply at any time without a mandatory waiting period.

The program arose from humble beginnings: "In 1954, when the U.S. Civil Commission [now the Office of Personnel Management] granted the postdocs status as two-year term federal employees, 21 scientists applied for the first competition, and NIST made six awards, paying $5,940 each. By 2004, the number of postdocs who may be hired annually is capped at 60 (Curry, 2004)." To date there have been over 6,000 applications to the NIST/NRC RAP and over 1,000 Research Associateships have been awarded.

THE ROLE OF THE NATIONAL ACADEMIES

The National Research Council—via its Fellowships Office—conducts the Research Associateship Programs in cooperation with sponsoring federal laboratories and research organizations approved for participation.[15] The Fellowships Office conducts a national competition to recommend and make awards (except NIST, which makes its own awards) to outstanding scientists and engineers at the postdoctoral level for tenure as guest researchers at participating laboratories.

---

[13] See Curry (2004) for biographies of selected NIST postdocs. NIST Fellows are esteemed senior scientists.

[14] Prior affiliation includes direct full-time employment relationships either with the laboratory or with a contractor whose work is performed there. A long-term consulting relationship usually makes an applicant ineligible.

[15] Material in this section was adapted from the NRC RAP website, "Associateship Programs at NIST," available at: http://qdev.boulder.nist.gov/817.03/jobs/nrc_local.htm. [Accessed June 24, 2007.]

The Associateship Programs office receives all application materials and supporting documents and conducts the competitive evaluation of applications. This evaluation for NIST/NRC Research Associateships (NIST/NRC RAs) is conducted by special panels convened for this purpose. Panelists are chosen to review applications on the basis of the applicants' stature and experience in the fields of science and engineering, and their evaluation becomes the basis from which awards are offered by NIST. Applicants are recommended for awards only after this open, national competition in which the panels rank candidates on the basis of quality alone. Final ranking in order of quality and the recommendation of applicants for awards are the exclusive prerogatives of the panels. Candidates for awards are selected by NIST from the panel's recommended list. NIST makes award offers throughout the spring and fall.

THE COMMITTEE'S CHARGE

The National Academies appointed an ad hoc committee to prepare this report (Appendix A). The committee's charge:

> "The Academic Competitiveness Council, in furtherance of the Administration's American Competitiveness Initiative, seeks to ensure that the nation invests wisely and effectively in educational programs to meet its science and technology goals. The ACC, therefore, requires evaluations of important STEM[16] education programs, including the NIST/NRC Postdoctoral Research Associateships Program. An ad hoc committee under the auspices of the Board on Higher Education and Workforce will describe the pool of applicants for and recipients of the NIST/NRC Postdoctoral Research Associateships and carry out a descriptive analysis of career outcomes of NIST postdoctoral scholars based on available information. As possible given available data, the committee will also describe how the program addresses agency goals. The committee will also outline an approach to evaluating the program relative to national S&E goals, NIST agency goals, and the value of the program to participants, which could be undertaken in a future study."

APPROACH AND SCOPE

To meet its charge, the committee focused on three objectives:

- To describe characteristics of NIST/NRC RAP applicants compared to the general pool of new science and engineering doctorates;
- To describe the experiences of NIST/NRC RAs compared to Research Associates in other programs; and
- To offer suggestions for conducting a more in-depth assessment of the careers of NIST/NRC RAs, with a particular focus on quantifying the benefits of the appointment to the recipients as well as to NIST.

The committee was guided by two principal questions:

---

[16] STEM is science, technology, engineering, and mathematics.

- Is NIST attracting the "best and the brightest" to the NIST/NRC RAP?
- What is the impact of the program on the Research Associates, NIST, and to the scientific and engineering communities in general?

In addition, the committee determined to offer recommendations where appropriate regarding data collection on applicants to the program, on the experiences of current and former NIST/NRC RAs, and on the views of NIST/NRC RAs and NIST employees toward the value of the program to NIST/NRC RAs, NIST, and to the scientific and engineering communities in general.

In assessing the NIST/NRC RAP, the committee sought to compare that program to the other RAPs administered by the NRC. First, the committee compared characteristics of applicants to the NIST/NRC RAP to other RAPs, and to a lesser degree to the general pool of recent science and engineering doctorates. Second, the committee compared characteristics and experiences of NIST/NRC RAs to Research Associates in other programs.[17] Finally, the committee undertook a very limited comparison of the careers of former NIST/NRC RAs to other former research associates, but data limitations prevented all but simple descriptive analyses.

The approach taken by the committee is limited in several ways. Most importantly, the analysis is not causal in nature. The committee cannot explain similarities or differences between NIST/NRC RAs and the comparison groups. However, description is an informative first step that reveals patterns and trends. It suggests potential causal hypotheses (e.g., the rise in the number of female applicants is due to the growing number of women receiving doctorates, better recruitment of women doctorates, or the changing nature of the profession, in which postdocs have become necessary in the career pathway) that can be tested in the future. Description also reveals gaps in data collection that might be addressed in the future. The analysis can also highlight similarities and differences between the NIST/NRC RAP and other RAPs, which can be useful in designing more rigorous evaluations to assess the impact of the award.

Second, the comparison between the NIST/NRC RAP and the other RAPs combined is imperfect. Not all of the RAPs have the same eligibility requirements (e.g., limited to U.S. citizens), nor are the appointments necessarily similar. However, in practice, these two groups of postdocs and postdoctoral appointments are quite close and there is the advantage that individuals in both groups fill out the same application form, awardees fill out the same final report, and Research Associate advisors fill out the same evaluation form—thus providing similar data.[18]

Third, the committee was limited by data availability (for details see below). Data do not exist back to the beginning of the NIST program in 1954. Comparative data on postdoctoral appointments and postdocs is relatively new—dating back to the mid- or late-1970s—and the

---

[17] Given the timeframe for completing the report, the committee was not able to also compare the applicants to the NIST/NRC RAs and those awarded appointments to comparable postdocs outside the RAPs, although such a comparison might be possible in the future with the cooperation of other agencies that maintain postdoctoral programs.

[18] Due to privacy concerns over the small number of individuals in some of the RAPs (because they do not make many awards or are relatively newer programs), it is not recommended to conduct analysis on each of the RAPs identified in Table 1-1.

most in-depth quantitative analysis of postdocs has only been undertaken in the past several years.

SOURCES OF INFORMATION

To complete its charge, the committee relied primarily on three sources of information: (1) data on applicants to the NIST/NRC and other RAPs and the experiences of NIST/NRC and other RAs, which are collected by the NRC's Fellowships Office; (2) data on the careers of former RAs collected by NIST and the Fellowships Office; and (3) data on S&E doctorates and postdoctorates collected by the National Science Foundation (NSF). In addition, the committee examined relevant literature on S&E postdoctorates, including surveys of postdoctorates, and examples of evaluations of other federal education programs. Finally, the committee conducted original research in the form of multiple expert panels with current and former NIST RAs and NIST employees. Table 1-2 summarizes the sources, which are further elaborated on below.

TABLE 1-2 Summary of Primary Data Sources

Topic	Data Source	General Focus	Coverage	Used in
Applications and awards	National Academies' "DataRAP" database[19]	Data collected from applications (especially demographic)	1965-2007	Chapter 2
	National Academies' "DataRAP" database	Question: how did you hear about the program?	1989-2007	Chapter 2
	NSF, Survey of Earned Doctorates	Demographic data and postgraduation plans of doctorates	1958-2005	Chapter 2
Experiences of Research Associates during the program	National Academies' "DataRAP" database	Demographic data (of applicants who were accepted)	1965-2007	Chapter 3
	National Academies' "DataRAP" database	Post-tenure exit survey of postdocs asking them about their experiences during the program.	1965-2006	Chapter 3
	National Academies' "DataRAP" database	Evaluation forms filled out by postdoc advisors	1965-2006	Chapter 3
Experiences of former Research Associates	National Academies' "DataRAP" database	Post-tenure exit survey of postdocs asking them about immediate career plans	1965-2006	Chapter 4
	NIST	List of former postdocs who stayed at NIST	1996-2007	Chapter 4
	Fellowships Office Directory[20]	List of former postdocs with current employer	1965-2002	Chapter 4

Data on applicants comes from two sources: a database constructed by the NRC's Fellowship Office to hold application data and the NSF's Survey of Earned Doctorates. The Fellowship's Office maintains a database on applicants, collecting information submitted by applicants during the application process. Until very recently, data were entered manually from an application form (see Appendix B); currently, some information is pulled from a web-based application. Data are available from 1959 (1966 for NIST/NRC RAs) through February 2007. Key variables are:

- Demographic information, including: birth date, citizenship, gender, race/ethnicity, and marital status;
- Educational background, including highest degree, degree field, and doctorate awarding institution;
- Desired postdoctoral position, including program and lab applied to;
- Outcome indicators, including the ranking of the applicants and the outcome of the applications; and
- How the applicant heard about the program.

---

[19] The DataRAP database coverage begins in 1959 for RAs other than NIST/NRC RAs, but as we are interested in comparing NIST to other federal RAs, it is appropriate to use the NIST/NRC parameters as the limiting factor.
[20] The Fellowships Directory coverage also begins in 1959.

In comparing NIST/NRC RA applicants to others who have recently received doctorates, one source of information is the NSF's Survey of Earned Doctorates (SED), which began in 1957-1958 to collect data annually on the number and characteristics of individuals receiving research doctoral degrees from all accredited U.S. institutions.[21] All individuals, as they receive their research doctorate, are asked to complete the survey. The SED collects information on the individual's education, characteristics, and postgraduation plans. Selected variables include:

- Academic institution of doctorate
- Birth year
- Citizenship status at graduation
- Country of birth and citizenship
- Field of degrees (N = 279)
- Marital status, number/age of dependents
- Postgraduation plans (work, postdoc, other study/training)
- Primary and secondary work activities
- Source and type of financial support for postdoctoral study/research
- Type and location of employer
- Race and Hispanic ethnicity (for selected sub-groups)
- Sex[22]

The survey has a greater than 90 percent response rate. Trend data are available back to 1957-1958. Partial data are available for the period 1920-1956.

Data on characteristics of those awarded research associates and their experiences during the postdoctoral appointment come primarily from two sources: the post-tenure survey given to research associates by the Fellowships Office and the NSF's Survey of Doctoral Recipients.

This information is supplemented by one-time surveys of postdocs conducted by The National Academies and by Sigma Xi.[23] Finally, the NSF's Survey of Graduate Students and Postdoctorates in Science and Engineering is also a potential resource.

The Fellowships Office conducts a post-tenure exit survey with individual research associates as they complete their postdoctoral appointment. The survey consists of a written questionnaire, and while nominally required, many postdocs do not fill it out. In addition, an evaluation form is requested of the research associates' advisors. Key variables from these two evaluations:

- Assessment of the research associate on innovative thinking, knowledge of field, motivation/initiative, overall science ability, independent research, and research technique;

---

[21] For more information, see: "SRS Survey of Earned Doctorates," available at: http://www.nsf.gov/statistics/showsrvy.cfm?srvy_CatID=2&srvy_Seri=1. [Accessed June 24, 2007.]

[22] The NSF uses the term "sex" instead of "gender." In this report, the terms "sex" and "gender" are used synonymously.

[23] Sigma Xi is "an international, multidisciplinary research society whose programs and activities promote the health of the scientific enterprise and honor scientific achievement. There are nearly 60,000 Sigma Xi members in more than 100 countries around the world. Sigma Xi chapters, more than 500 in all, can be found at colleges and universities, industrial research centers and government laboratories." From "About Sigma Xi," available at: http://www.sigmaxi.org/about/overview/index.shtml.

- Productivity during postdoctoral tenure, including: number of domestic and international presentations given, number of peer-reviewed journal articles, number of patents applied for, and number of awards received;
- Research associate' views of the program, including: quality of mentoring from the Lab NRC Advisor, quality of laboratory support (e.g., equipment), how the National Academies Associateship award affected your career to date, quality of administrative support from the Laboratory, development of knowledge, skills, and research productivity, and evaluation of advisor (open ended); and
- Future plans, including: future plans of the research associate, level of post tenure position, organization with whom post-tenure associate will be affiliated, and title of post tenure position.

The principal source for information on postdocs in general—particularly outside academia—is the NSF's Survey of Doctoral Recipients (SDR).[24] The SDR collects information from individuals who have obtained a doctoral degree in a science, engineering, or health field. The SDR is biennial, and data are available for 1973-2003.[25] The SDR is a longitudinal survey that follows recipients of research doctorates from U.S. institutions, who were living in the United States during the survey reference week, and who are under age 76.

Using postdoctoral status as a starting point, a number of potential relevant variables can be assessed. Selected variables include:

- Citizenship status
- Country of birth
- Country of citizenship
- Date of birth
- Educational history (for each degree held: field, level, institution, when received)
- Employment status (unemployed, employed part time, or employed full time)
- Geographic place of employment
- Marital status
- Number of children
- Occupation (current or past job)
- Primary work activity (e.g., teaching, basic research, etc.)
- Postdoctorate status
- Publication and patent activities
- Race/ethnicity
- Salary
- Satisfaction and importance of various aspects of job
- Sector of employment (e.g., academia, industry, government, etc.)
- Sex

---

[24] For more information on the SDR, see: "SRS Survey of Doctoral Recipients" available at http://www.nsf.gov/statistics/showsrvy.cfm?srvy_CatID=3&srvy_Seri=5. [Accessed June 24, 2007.]

[25] The next SDR will cover 2006 and future surveys are expected to switch to even-numbered years.

As noted by the NSF: "There have been a number of changes in the definition of the population surveyed over time. For example, prior to 1991, the survey included some individuals who had received doctoral degrees in fields outside of S&E or had received their degrees from non-U.S. universities. Because coverage of these individuals had declined over time, the decision was made to delete them beginning with the 1991 survey. Survey improvements made in 1993 are sufficiently great that SRS staff suggest that trend analyses between the data from the surveys after 1991 and the surveys in prior years must be performed very cautiously, if at all."[26] A more limiting factor for the committee was that the survey merely asks respondents if they are currently in a postdoctoral position—but does not specify how long. Therefore, some respondents could be just starting a postdoctoral appointment (probably their first, but not necessarily) while others could be in the first, second, or third year of a position. This makes it difficult to compare the experiences of postdocs identified in the survey with research associates.

In the most recent wave of the SDR—collected in 2006, though not available at the time this report was prepared—a series of questions relating to postdocs was added. Questions include: whether the respondent is currently employed as a postdoc and the number of postdoctoral appointments the individual has had. For each appointment, respondents are asked to provide data on start and end dates, the primary reason for taking the position, employment sector (e.g., academic, industry, government), whether the employer provided health or retirement benefits, and to what extent the most recent or current postdoctoral appointment impacted on the respondent (e.g., increasing knowledge, research skills, career opportunities). If these questions continue to be asked in future SDRs, these data would be very helpful in getting a broad picture of postdocs' experiences.

A one-time survey of postdocs was conducted by Sigma Xi, in partnership with the National Postdoctoral Association, the American Association for the Advancement of Science, and the National Bureau of Economic Research, in 2003. Invitations to participate in the survey were sent to 1,432 provosts and vice provosts, deans, human resources personnel, and leaders of postdoc offices and associations at 174 institutions. Of these, 46 institutions, including 18 of the top 20 academic employers and the largest government employer of postdocs, the National Institutes of Health (NIH), agreed to participate (Davis, 2005). Email was used to contact the 22,000 postdocs at the participating institutions, roughly 40 percent of all postdocs working in the United States at the time. The survey's response rate was 34 percent.

The survey focused on the following characteristics of postdocs:

- Demographic information (e.g., sex, race/ethnicity, age, citizenship);
- Benefits and services (e.g., salaries, benefits, and the adequacy of various resources);
- Institutional environment (written policies and procedures and about the availability of various types of training);
- The postdoc's advisor;
- The postdoc's position;
- Outcomes (e.g., time allocation, papers);
- Career plans; and
- The postdoc's satisfaction with the postdoctoral appointment.

---

[26] "SRS Survey of Doctoral Recipients" available at http://www.nsf.gov/statistics/showsrvy.cfm?srvy_CatID=3&srvy_Seri=5. [Accessed June 24, 2007.]

A departmental survey by the NSF is the Survey of Graduate Students and Postdoctorates in Science and Engineering (GSS).[27] The survey includes data on the number and characteristics of postdocs in S&E fields in U.S. institutions. The survey is conducted annually, and it is targeted to individual academic departments at graduate institutions. Data include count data on postdoctorates by source of support, sex, and citizenship. Data was collected beginning in 1966, and the most current data, as of this writing, is for 2005. However, the collection of data on postdocs has changed over time, making it difficult to examine long term trends.[28]

Finally, a one-time survey of institutions with postdocs was conducted by the Committee on Science, Engineering, and Public Policy (COSEPUP) at The National Academies. "COSEPUP decided to survey the top 25 academic institutions (in terms of the largest numbers of postdoctoral scholars) and five each of the following: smaller institutions (in terms of number of postdoctoral scholars), medical schools, historically black colleges and universities (HBCUs), industry, research institutions, and government laboratories. The survey was conducted from November 1999 to April 2000. The survey was conducted of 49 organizations who have postdoctoral scholars. Forty of the 49 organizations responded (82 percent response rate) (NAS/NAE/IOM, 2000:138)."

Career data are more difficult to come by. One source of information is The National Academies' Associateships Program Directory: "A Directory of former NRC Associates was published in book form in 1995. Since then, more than 2,000 additional associateships have been awarded. In this web Directory, we have updated current information on many of our former Associates. This information can be viewed by Associate name, program, or date of award, or by state or country of the Associate's current employment. The Directory is fully searchable."[29] Data cover 1959-2002. Using these data, it is possible to see where former NIST and other research associates have ended up—with the caveat that the information is limited to their current position, when they were asked to respond.

A second source of career data consists of monthly reports on NIST Research Associates collected and maintained by NIST. These data record changes in the employment status of NIST Research Associates. NIST Research Associates may resign or be terminated (although they may then work as contractors with NIST) or stay at NIST by being converted to a term position or by being converted to a career conditional position. Data are available in hardcopy for 1998-2001 and in electronic format for 2002-2007. Since these data cover research associates at the end of their tenure; these data cover research associates who began their appointments in 1996 or later.

A third source of information on the careers of postdocs which might be helpful comes from the NSF's SDR. It is possible to examine respondents who completed multiple surveys, so one could identify survey respondents who were postdocs in the 1999 survey and examine changes in employment or other characteristics, for those respondents who also responded to the 2001 survey. The NSF is currently in the middle of a feasibility study designed to explore ways to fill

---

[27] For more information on the survey, see: "SRS Survey of Graduate Students and Postdoctorates in Science and Engineering," available at: http://www.nsf.gov/statistics/showsrvy.cfm?srvy_CatID=2&srvy_Seri=2. [Accessed on June 24, 2007.]

[28] Of particular note is that prior to 1975, NSF did not seek to collect data on all doctorate-granting institutions. For a fuller description of changes in the survey over time, see "SRS Survey of Graduate Students and Postdoctorates in Science and Engineering," available at: http://www.nsf.gov/statistics/showsrvy.cfm?srvy_CatID=2&srvy_Seri=2. [Accessed on June 24, 2007.]

[29] NRC Associateships Program Directory, available at: http://nrc58.nas.edu/pgasurvey/data/aodir/gen_page.asp. [Accessed on July 13, 2007.]

in the gap in data on postdocs. One goal is to improve coverage in two ways: by obtaining information on more postdocs (e.g., foreign Ph.D.-degreed postdocs or M.D.-Ph.D. postdocs) than might be covered in the SDR; and by covering more sectors beyond academia (e.g., postdocs at non-profits or government agencies), which fall outside the scope of the GSS. The NSF's study may lead to a future survey of postdocs.

Finally, the committee undertook original data collection in the form of three expert group discussions with current NIST RAs (CRA), former NIST RAs still working at NIST (FRA), and NIST RA advisors and division chiefs (AD). Each of the panels lasted approximately one hour and occurred on July 25, 2007. Questions asked of these participants (with the person the questions were directed to in parentheses) are listed in Box 1-1.

Nine current research associates participated in the first expert panel. Six were male. Three began their appointments in 2005, five began in 2006, and one began in 2007. The research associates represented seven different labs. Six former research associates still at NIST participated in the second expert panel. Four were male. One finished their postdoctoral appointment in 2005; four finished in 2006, and one had just finished in 2007. Three had been converted to career conditional status; the other three had converted to term appointments. Participants represented five labs. The third panel consisted of eight advisors and lab chiefs. Two had been former NIST RAs. Seven were male. They represented five labs.

Box 1-1
Expert Panel Questions

- What do you see as the value of the NRC/NIST postdoctoral program? (AD)
  - To the postdocs
  - To the NIST mission?
  - To your science/engineering field?
- Is there any downside to the program? (AD/FRA)
- Are there areas where you think it could be improved? (AD/FRA)
- In what ways have you benefitted from being an NRC/NIST postdoc? (FRA)
- Do the postdocs differ in any way from other entry level NIST employees? (AD)
- What effect do you think the postdoctoral program has on the careers of NIST postdocs? (AD)
- What effect do you think the postdoctoral program will have on your career? (CRA/FRA?)
- Has the postdoctoral experience been what you expected ?(If not, explain.) (CRA)
- Would you recommend changing anything about the postdoctoral program? (AD /CRA/FRA)
- Why did you apply for the NIST postdoctoral program? (CRA/FRA)
- What other postdoc programs, jobs, and other things did you apply for? (CRA/FRA)
- Has having a NIST postdoc affected where you might like to be employed? (CRA)
- What are your career plans after the postdoc ends? (CRA)
  - Has this changed as a result of your experience in the NIST postdoc?
- Do you think NIST postdocs differ from other postdocs in science and engineering? (AD)
- Are you satisfied with the current selection process for awarding NIST postdocs? (everyone)
- If this postdoc program did not exist, how would that affect your work and staff?AD)
- Are there any NIST postdocs that stand out in your mind as having made a significant impact on the field? (AD)
- Do you continue to have contact with former postdocs (in what ways)? (AD)
- Is there anything else you would like to tell us about the program?(everyone)

OUTLINE OF REPORT

The report is organized in a chronological way, following the trajectory of doctorates from application, to award, to their experiences in the Program, and concluding with a career assessment of awardees after their appointments have ended. Chapter Two examines applicants to the NIST/NRC RAP and compares them to applicants to other RAPs. It also examines applications and awards disaggregated across several dimensions, such as gender or doctoral-granting institution. Chapter Three examines the experiences of NIST RAs and RAs at other federal agencies, as well as research associates' views on the value of the Program. Chapter Four examines the careers of former research associates. Chapter Five presents an overall summary of preliminary results and recommendations.

2

Recruitment and Selection

The NRC Research Associateship Program at NIST (hereafter "NIST/NRC RAP") is thought to be a value to both the postdoctoral recipients and to NIST itself. This chapter is divided into three sections. First, the recruitment of RAs is put into context by examining trends in Ph.D. production and trends in postdoctoral appointments. Second, the application process is examined. Finally, characteristics of applicants and awardees are described.

TRENDS IN DOCTORATES AND POSTDOCTORATES

Trends in Doctorates

As noted in the previous chapter, information on doctorates comes from the National Science Foundation's Survey of Earned Doctorates (see Appendix B for a recent questionnaire). The number of doctorates granted in the United States has generally grown over the past 100 years, peaking in 1973 and 1998. Prior to 1953, more doctorates were produced at private institutions (NSF, 2006c). Examining trends from 1920-1999, for all doctorates—not just U.S. citizens:

- About two-thirds of doctorates were awarded in science and engineering (S&E).
- Between 1920 and 1974, between 87.2 and 93.6 percent of doctorates in S&E were awarded to men. This figure dropped steadily from 1975 to 1999. In the period 1995-1999, it stood at 66.8 percent.[30]
- A growing percentage of S&E doctorates were awarded to foreign nationals: 38.6 percent by the 1990s.[31]
- Underrepresented minorities receive few Ph.D.s relative to whites and Asians: 7.4 percent of S&E doctorates awarded in the second half of the 1990s went to underrepresented minorities.[32] But the number and percentage of underrepresented minorities receiving S&E Ph.D.s has grown from the 1970s to the 1990s.
- The median age of doctorate recipients has been increasing over time; although the median age for recipients of Ph.D.s in S&E is much lower than the age for those receiving non-S&E Ph.D. degrees.
- A majority of doctoral recipients were married at the time of graduation, though the percentage of married graduates has been declining since the 1960s (NSF, 2006c).

For the years 2000 to 2005, some trends have continued (see appendix C for underlying data). In 2005, S&E doctorates accounted for 64 percent of all doctorates awarded, which is similar to the ratio in the 1990s. However, more and more women are receiving doctorates in S&E. In 2005, about 38 percent of S&E Ph.D.s went to women. The number of U.S. citizens

---

[30] These data are available at: http://www.nsf.gov/statistics/nsf06319/pdf/fig03-03.pdf.

[31] These data are available at: http://www.nsf.gov/statistics/nsf06319/pdf/fig03-06.pdf.

[32] Underrepresented minorities includes American Indian/Alaskan Native, Black, and Hispanic; and excludes Asian/Pacific Islander.

receiving doctorates in science and engineering has declined somewhat from 2000 to 2005. Among doctorates where citizenship was known, in 2005, only about 56 percent of S&E Ph.D.s were awarded to U.S. citizens. Finally, the number of American Indian/Alaska Natives receiving S&E Ph.D.s has declined from 2000 to 2005; the number of Black/African Americans receiving S&E Ph.D.s has stagnated; and the number of Hispanics receiving S&E Ph.D.s has increased somewhat. As a result the same percent—about 10—of S&E Ph.D.s went to underrepresented minorities.

Trends in Postdoctoral Appointments

Postdoctoral appointments date back over 100 years; however the hiring of postdocs did not grow significantly until the second half of the twentieth century. An initial period of rapid growth occurred in the 1950s, stimulated by the Cold War demand for scientists and engineers. In the 1970s, and again during the recession of the 1990s, the number of postdoctoral positions increased due to a weaker economic market for Ph.D.s. (NAS/NAE/IOM, 2000; Davis, 2005).

Postdoctoral appointments can provide benefits both to the recipients and the employers. For postdocs, the position is a way to obtain further training. Postdoctoral appointments in federal labs or industry can be an entrée into non-academic careers. Concerning the impact on the employer, one report notes that "As a whole, the postdoctoral population has become indispensable to the science and engineering enterprise, performing a substantial portion of the nation's research in every setting. For example, a survey of research articles in two recent issues of *Science* found that 43 percent of the first authors were postdocs.[33] In many labs, postdocs also educate, train, and supervise junior members, help write grant proposals and papers, and present the laboratory's research results at professional society meetings" (NAS/NAE/IOM, 2000:10). However, it is important to note that there have been some complaints about the situation for postdocs.

According to the NSF, in 2005 there were approximately 35,000 postdocs in academia, across all science and engineering fields broadly defined (NSF, 2007). However, there are differences by field. "In some fields, such as computer science and engineering, there is relatively little incentive to pursue a postdoc—or even a Ph.D.—because rewarding jobs are available at the bachelor's and master's levels. In other fields, such as biology and physics, a postdoc is virtually mandatory, especially for academic employment" (NAS/NAE/IOM, 2000:14). Table 2-1 gives a field breakdown for number of postdoctoral appointees, while Table 2-2 lists the percentage of doctoral recipients with definite plans to pursue postdoctoral study or research by field.

---

[33] Vogel, G. *Science*, 1999, Vol. 285, p. 1531.

TABLE 2-1 Science and Engineering Postdoctoral Appointees in Doctorate-Granting Institutions, by Field, 1998-2005

Field	1998	1999	2000	2001	2002	2003	2004		2005
Science and engineering	27,826	28,943	30,197	30,163	31,871	33,516	33,898	r	34,535
Science	24,973	25,747	26,884	26,997	28,303	29,696	29,935	r	30,374
Agricultural sciences	695	749	822	835	945	1,052	941		988
Biological sciences	15,755	16,091	16,729	17,022	17,640	18,605	18,675	r	18,995
Computer sciences	371	332	341	335	359	358	384		406
Earth, atmospheric, and ocean sciences	898	923	1,155	1,036	1,113	1,166	1,253		1,364
Mathematical sciences	279	351	385	353	391	447	466		496
Physical sciences	5,973	6,136	6,252	6,198	6,587	6,707	6,945		6,865
Engineering	2,853	3,196	3,313	3,166	3,568	3,820	3,963		4,161

Note: r = data significantly revised; replaces previously published data.

Source: National Science Foundation/Division of Science Resources Statistics, Survey of Graduate Students and Postdoctorates in Science and Engineering in NSF (2007). Adapted from Table 49.

TABLE 2-2 Percent of Doctoral Recipients with Definite Commitments Who Plan Postdoctoral Study or Research, by Broad Field of Study, 1982 and 2002

Field of Study	1982	2002
Biological sciences	72.1	74.4
Physics/astronomy	48.8	66.7
Chemistry	39.8	52.9
Earth, atmospheric, ocean sciences	25.9	51.6
Mathematics	15.8	42.5
Agricultural sciences	15.9	38.3
Engineering	11.4	24.8
Health sciences	15.4	21.1
Computer sciences	9.1	19.7

Source: NSF/NIH/USED/NEH/USDA/NASA, Survey of Earned Doctorates, in Hill et al., 2004: Figure 1.

As there are differences by field, so too do postdocs vary by demographic characteristics. Based on the Sigma Xi survey of postdocs, the following conclusions are noted:

- The majority of postdocs in the life and health sciences, in the physical sciences, and in engineering are men. Men also comprise the majority of postdocs who are temporary-visa holders.
- About 75 percent of citizen and permanent resident postdocs identified themselves as white.
- The majority of postdocs responding to the survey held temporary visas. 40 percent were U.S. citizens and 6 percent were permanent residents.
- The majority of postdocs were between 30 and 35 years old; 69 percent were married or otherwise partnered; and about a third had children (Davis, 2005).[34]

[34] It is difficult to know how generalizable the results of this survey are: the percentages are based on a 34 percent response rate and from postdocs at select institutions. Nonresponse bias may have affected the survey estimates.

THE SELECTION PROCESS

The process by which applicants apply and are selected to become NIST/NRC RAs can be summarized in a few basic steps:

- Potential applicants hear about the NIST/NRC RAPs
- Applicants apply to the Program
- Applications are reviewed by expert panels overseen by the NRC's Fellowships Office. Each applicant receives a rating based on the average scores of three reviewers (possibly two more if the scores are disparate (differ by 1.5 points between highest and lowest score)
- The Fellowships Office forwards ranked (highest to lowest rated) applicants on to NIST
- Partly on the basis of those rankings, NIST offers selected applicants postdoctoral positions
- Most of those who receive offers accept and become research associates

There are a number of ways to publicize postdoctoral positions. Both the National Academies and NIST have links to the program on their respective websites.[35] Staff from the National Academies attend conferences where they make information on the program available. Advertisements are also placed in relevant publications, such as *Physics Today*, *Science*, and the *Chronicle of Higher Education.*

Once prospective applicants hear about the program, the next step is to complete an application (see Appendix D). Noteworthy information collected by the application includes:

- Educational background
- Demographic data (e.g., citizenship, gender, date of birth, race/ethnicity, marital status)
- How the applicant heard about the program
- Previous research and publications
- A research proposal intended to be carried out during the postdoctoral tenure

Applicants also have recommendations submitted on their behalf. In addition, the proposed research advisor at the laboratory/center also reviews the applicant's proposed research project.

Completed applications are collected by the National Academies. This is followed by a review process, as described on the National Academies' Web site "Review Criteria" (see Box 2-1).[36]

---

[35] The National Academies, "RAP Home," available at: http://www7.nationalacademies.org/rap/ and NIST, "NIST Postdoctoral Research Associateships Program," available at: http://www.nist.gov/oiaa/postdoc.htm.

[36] The National Academies, "Review Criteria," available at: http://www7.nationalacademies.org/rap/Review_Criteria.html.

Box 2-1
Review Criteria

Applications for awards from the NRC Research Associateship Programs are reviewed by panels of experts in 6 broad discipline areas: Chemistry; Earth and Atmospheric Sciences; Engineering, Applied Sciences and Mathematics; Life Sciences; Physics; and Space Sciences. Each application is read by a minimum of 3 panelists. Panelists assess the quality of an application, the likelihood for success and the contribution of the research to the mission of the sponsoring federal laboratory. Postdoctoral applicants are evaluated on the basis of demonstrated ability as a student and on their potential for making contributions as an independent scientist. Senior applicants, including applicants to Summer Faculty programs, are evaluated on the basis of proven ability and demonstrated research accomplishments. Evaluations are made without regard to age, sex, marital status, national origin, creed, racial group, or ethnic group.

Each application is assigned a numerical score and the applicant's final score is an average of all reviews. Scoring is on a 10-point scale and only applicants scoring 7.5 or above are considered for awards. Sponsoring laboratories offer awards to the highest scoring applicants first and continue to make awards until available slots are filled. In the review process each applicant is evaluated on four major elements with the approximate weighting as indicated:

Scientific merit of the proposed research (40%)

The research proposal is the most important element of the application and as such is weighted most heavily in the review. The proposal is evaluated for: importance of the proposed research area, clearly stated objectives, technical soundness of the work plan, innovative aspects of the proposal, feasibility of success, timeliness (can the proposal be completed in the allotted time), likelihood that the research will result in publication, and contribution of the research to the mission of the sponsoring laboratory.

Reference reports or letters of recommendation (20%)

Reference Reports or letters of reference contain opinions of persons who should have had a close professional relationship with an applicant; references provide reviewers with important information regarding the applicant's scholarly abilities. Reference reports are given greater weight for Postdoctoral applicants, where a publication record may not be as extensive as that of a Senior applicant.

Academic and research record (20%)

Panelists review the appropriateness of the applicant's training for the proposed research, previous research experience and record of publication. For Postdoctoral applicants only, a transcript of the academic record is required.

Laboratory technical evaluation (20%)

The Laboratory/Center Review form includes comments of the prospective Advisor and the disposition of the Laboratory/Center's program committee or representative concerning the suitability of the applicant's proposed research. This information aids reviewers in determining the value of the proposed research to the sponsoring agency.

During the expert panel meetings, applicants are rated from 10 to 0, with 10 being the highest possible score. In practice, scores have ranged from 9.90 to 0 and applicants' scores may differ by as little as 0.02 (e.g., 9.65 to 9.63). The review process is seen as something of a mystery to NIST staff. During the expert panel with former research associates, they commented that they did not understand how candidates were ranked. Former research associates had their own ideas about how candidates should be ranked. As a consequence of the personal connections between NIST employees and potential applicants, NIST employees formed opinions about who they thought top candidates were, and then registered their surprise when those people did not end up ranked at the top. In a related comment, former research associates commented that they would like to have a much better idea of how the research project that the applicant intends to work on (as described in the application) is graded. They felt that the current grading system is not transparent and is too general. NIST staff submit a recommendation on behalf of applicants and the staff do not seem sure what they ought to stress in their recommendation.

Advisors and division chiefs focused on another issue in their expert panel: the review panels. They questioned whether the NRC review panels were organized well. They were concerned the panels were skewed to reviewers from academia. They did not know who serves on the panels. They would like to have more input into panel makeup, for example by suggesting names of potential reviewers (and then knowing if their suggestions were used). Another concern was whether rankings were normalized across panels.

The applicants' reviews, with scores, are sent to NIST, which selects the applicants to be offered postdoctoral positions. In practice, most of those offered, will accept a Research Associate.

RECRUITMENT

A concern for NIST and The National Academies is how well the program is reaching out to potential applicants. During the expert panels, current research associates noted that NIST was often their first choice. Reasons why included: family reasons, quality of advisors, ability to collaborate, and location. The research associates had applied to multiple positions, for example in academia or other government labs. Concerning how research associates heard about the position, the most frequent answer involved personal communication—either they met their future advisor at a conference, job fair, or when the advisor gave a lecture at their school; or a graduate advisor suggested they apply. (Personal communication was also repeatedly mentioned by the advisors and chiefs—that personal relationships had the best return on investment.) In several cases, their advisors were former research associates. Word of mouth was very important.

Former research associates echoed these comments. Answers given by participants included: had a personal relationship—in one case the former research associate's spouse worked at NIST; were recruited by advisors (at conferences or presentations made by NIST employees at universities); met researchers at NIST (this was suggested by former RA from local universities); and came across the Web site describing the RAPs. One RA had applied to a regular position at NIST and then saw the listing for the RAP. The former RAs noted that sometimes staff invite graduate students to give research talks at NIST as a way to bring potential applicants out to NIST. It did seem to the participants that the personal connections were much more effective. One former research associate noted that there is less outreach by NIST in the biological areas. This is important, as NIST may be moving in the future to more interdisciplinary research that has a biological component and there will be a need for more people with training in biological sciences. A physicist noted that at one point in the past, there was an advertisement in *Physics Today*, but he had not seen it recently. He felt that a better job of advertising could be done. Finally, former research associates noted that some had applied to multiple postdoctoral positions and had chosen NIST as the better alternative.

Advisors and lab chiefs did note that in some areas they felt that the number of applications was low. Of particular concern were some areas within engineering and computer science. They noted that doctorates had many options in these areas and many forego postdocs, as well as the view that there are more foreign nationals and fewer U.S. citizens getting degrees in these areas. They did agree with the current and former research associates that personal communication seemed to be the best way to recruit. Participants in this panel did note that different labs differed in how they attempted to recruit applicants and to what degree they tried.

Overall, participants reported that:

- The program is very prestigious (although some current RAs felt that the program was less so);
- The program is well-known (again, with some minority comment that it is not that well known); and
- The most common way people heard about the program was through personal communication.

To examine these findings more broadly, the committee turned to data collected on the NRC's Fellowships Office RAP application form. The application for postdoctoral programs includes a question on how the applicant heard about the position.[37] Applicants were requested to select one of the following: colleague or fellow graduate student; Ph.D. thesis advisor or other professor; university placement office; former or current NRC Research Associate; research advisor or other scientific staff at the federal Laboratory; RAP's staff member at professional scientific meeting; Advertisement in professional publication; or other. Data are available for 1989 to 2007.

The dataset contained 24,849 applications, of which 2,743 were applications to the NIST/NRC RAP. The number of applicants is much less than this since applicants can apply for multiple positions in the same year or across years. (Because the NIST/NRC RAP was until recently reviewed once per year, there are only a handful of cases where an applicant applied more than once per year to this particular program.)[38] In assessing how applicants heard about the program, we combined information from individuals who applied for multiple positions, if they selected different information sources, into a single record. This was done because in most cases, an applicant applying to multiple positions identified the same source in each case. Thus, the dataset was updated so that there was one record for each individual, regardless of how many applications they submitted. An exception occurred for applicants who applied to both the NIST/NRC RAP and any other RAP. Since a goal is to compare the sources identified by applicants to NIST and all other federal agency RAP postdoctoral positions, any individual who applied to both programs remained in the database twice. Two hundred and thirty-six individuals who applied to both NIST and other federal positions at the same time fit this

---

[37] The question is: "To assist us in making information available to a greater number of potential applicants, it is important for us to learn how you initially heard about the National Academies RAPs."

[38] It is not clear if this is a data entry error.

exception. After reducing duplicate entries, the database consisted of 12,737 records: 2,717 applicants to NIST and 10,020 applicants to other federal RAPs (with 236 applicants appearing on both lists). The NIST applicants identified a total of 2,890 sources and the non-NIST applicants identified a total of 10,497 sources.

As Table 2-3 shows, the applicants to the NIST/NRC RAP were twice as likely as applicants to the other RAPs to hear about the position initially from their Ph.D. advisor or other professor and somewhat more likely to hear about the program from colleagues or fellow graduate students. Surprisingly, they were less likely to hear about the program from a research advisor or other scientific staff at the federal laboratory, compared with applicants to other federal RAPs.

TABLE 2-3 How Applicants First Heard About the Research Associateship Program, 1989-2007

Source of Information	NIST/NRC RAP (%)	Other RAPs (%)	All Programs (%)
Colleague	23.3	20.9	21.4
Professor	33.7	16.6	20.3
Placement office	1.0	1.7	1.5
NRC associate	8.2	9.3	9.1
Laboratory staff	18.0	26.7	24.8
Journal	3.0	8.7	7.4
NRC staff	0.2	0.3	0.3
Professional meeting	4.8	5.3	5.2
Other	7.8	10.6	10.0

Source: National Academies, DataRAP Database, tabulations by staff.

Regarding advertisements in professional publications, applicants to NIST/NRC RAP identified about 30 publications, while applicants to non-NIST RAPs identified approximately 190 publications. (However, since the other RAPs include a broader set of fields, this larger number of publications should be expected and the smaller number of publications for the NIST/NRC RAP is not an indication, by itself, of less effort to reach potential applicants via publications. Top publications identified by applicants to NIST/NRC RAP: *Physics Today*, Fellowships Office mailing, *Chemical & Engineering News*, *Mechanical Engineering*, *Spectrum of the IEEE*, and *Science*. Examining trends in the sources cited over time, as is done in Table 2-4, shows that applicants for NIST/NRC RAs did not usually first find out about the program via an advertisement in a publication.

TABLE 2-4 How Applicants to the NIST/NRC Research Associateship Program First Heard About the Program, 1989-2007

Year	Colleague (%)	Professor (%)	Placement Office (%)	NRC Associate (%)	Lab Staff (%)	Journal (%)	NRC Staff (%)	Professional Meeting (%)	Other (%)	N
1989	100.0	0.0	0.0	0.0	0.0	0.0	0.0	0.0	0.0	1
1990	10.3	38.3	0.9	5.6	27.1	6.5	0.0	1.9	9.3	107
1991	33.3	63.0	3.7	13.9	27.8	12.0	0.0	9.3	18.5	108
1992	22.5	34.7	2.3	10.8	20.3	5.0	0.0	3.2	1.4	222
1993	22.6	35.5	0.9	9.0	15.0	6.8	0.9	5.1	4.3	234
1994	15.6	18.9	0.9	5.3	5.0	2.9	0.2	2.6	2.0	456
1995	34.3	37.7	1.0	0.5	17.4	1.9	0.0	4.8	2.4	207
1996	31.1	28.8	0.8	0.8	24.2	2.3	0.0	6.1	6.1	132
1997	15.6	14.7	0.6	3.2	6.5	0.9	0.3	2.4	6.8	339
1998	38.2	27.3	0.0	4.5	14.5	0.0	0.9	5.5	9.1	110
1999	23.3	22.5	0.8	14.2	23.3	0.8	0.0	2.5	12.5	120
2000	6.1	15.7	0.9	6.1	12.2	0.9	0.4	1.7	5.7	230
2001	21.7	34.8	0.0	13.0	19.6	0.0	0.0	4.3	6.5	46
2002	18.1	23.3	0.0	14.4	23.3	1.9	0.0	8.4	10.7	215
2003	11.5	24.5	0.0	4.2	10.3	0.8	0.0	6.1	6.9	261
2004	20.3	38.0	0.6	7.6	17.7	1.3	0.0	1.3	13.3	158
2005	23.1	34.0	1.3	12.2	16.7	1.3	0.0	2.6	9.0	156
2006	10.2	23.6	0.6	3.8	12.7	0.6	0.0	3.2	2.9	314
2007	22.4	40.2	0.0	6.5	15.9	0.9	0.0	4.7	9.3	107

Source: National Academies, DataRAP Database, tabulations by staff.

Journals seemed to be more important as a source in the first half of the 1990s than in the current decade. Another finding of note is that placement offices at universities tend to be an infrequently cited source of information. Although the committee did not have prior expectations, it was still a bit surprising that the proportion of applicants hearing about the program from former or current RAs was not that great.

We next examined some characteristics of applicants to the NIST/NRC RAP to see if different types of applicants differed on how they first heard about the program. Comparing men and women, we found no significant differences, except for presentations at professional meetings, where women were twice as likely as men to first hear about the program.

TABLE 2-5 How Applicants to the NIST/NRC Research Associateship Program First Heard About the Program, by Gender, 1989-2007

Source of Information	Women (%)	Men (%)
Colleague	22.2	23.6
Professor	32.1	34.1
Placement office	0.6	1.1
NRC associate	8.1	8.2
Laboratory staff	17.4	18.2
Journal	2.4	3.1
NRC staff	0.2	0.2
Professional meeting	8.3	4.1
Other	8.7	7.6
N	505	2385

Source: National Academies, DataRAP Database, tabulations by staff.

Turning to race/ethnicity, we compared how whites and all other ethnic/racial groups first heard about the program. Similar to women, the results were fairly consistent across these two groups and minorities were more likely than whites to hear about the program via a presentation at a professional meeting—but not significantly so.

TABLE 2-6 How Applicants to the NIST/NRC Research Associateship Program First Heard about the Program, by Race/Ethnicity, 1989-2007

Source of Information	White (%)	All Other (%)
Colleague	23.2	20.7
Professor	34.6	33.0
Placement office	0.9	1.1
NRC associate	8.0	8.6
Laboratory staff	18.0	18.7
Journal	2.5	2.9
NRC staff	0.1	0.6
Professional meeting	4.9	6.6
Other	7.9	7.8
N	2298	348

Source: National Academies, DataRAP Database, tabulations by staff.

As a next step, future analysis could examine the relationship between different sources of information among applicants and outcomes of applications. For example, Table 2-7 examines this association in general for applicants to NIST/NRC RAPs and other RAPs.

TABLE 2-7 Percent of Awardees Among Applicants by Source of Information About the Program, 1965-2007

Source of Information	Awardees among Applicants (%)	
	NIST/NRC RAP	Other RAP
Colleague	28.2	36.1
Professor	26.1	40.3
Placement office	15.6	30.5
NRC associate	31.4	43.2
Laboratory staff	28.3	47.3
Journal	17.2	26.4
NRC staff	100.0	42.3
Professional meeting	28.4	37.0
Other	29.2	28.8

Source: National Academies, DataRAP Database, tabulations by staff.

As Table 2-7 illustrates, successful applicants to the NIST/NRC RAP were more likely to hear about the Program from NRC staff or an NRC associate; although in these cases, few applicants had heard about the Program from these sources. There seems to be much more variability in how successful applicants hear about the NIST/NRC RAP, as compared with successful applicants to the other RAPs. Table 2-8 focuses on just those applicants who received RAs.

TABLE 2-8 Percent of Awardees by Source of Information About the Program, 1965-2007

Source of Information	Awardees (%)	
	NIST/NRC RAP	Other RAP
Colleague	23.9	19.4
Professor	32.2	17.0
Placement office	0.6	1.3
NRC associate	9.3	10.4
Laboratory staff	18.3	32.7
Journal	1.9	5.9
NRC staff	0.2	0.3
Professional meeting	5.1	5.0
Other	8.4	7.9
N	825	4022

Source: National Academies, DataRAP Database, tabulations by staff.

Table 2-8 shows that about one-third of awardees to the NIST/NRC RAP first heard about the program via a professor; whereas about one-third of awardees to the other RAPs first heard about the program from lab staff. Taken together, Tables 2-6 to 2-8 suggest that personnel communication was the most important mechanism for transmitting information about the pograms to prospective applicants. This sort of analysis could be extended by focusing on subsets of applicants, that is, by gender, race/ethnicity, or discipline, to see how well outreach succeeds.

PRELIMINARY RESULTS

First, the application form is a very useful data collection instrument. Among the three current instruments—application form, final report, adviser's evaluation—the application form has produced the most data. Second, personal communication is the most likely means by which applicants hear about the RAPs, including the NIST/NRC RAP. Key findings regarding how applicants heard about the program:

- Applicants to the NIST/NRC RAP were twice as likely as applicants to the other RAPs to hear about the position initially from their Ph.D. advisor or other professor and somewhat more likely to hear about the program from colleagues or fellow graduate students, but less likely to hear about the program from a research advisor or other scientific staff at the federal laboratory;
- The most common sources of information for applicants to the NIST/NRC RAP were professors or colleagues;
- Male and female applicants heard about the NIST/NRC RAP similarly, except via presentations at professional meetings, which women cited twice as often as men; and
- There were no differences by race/ethnicity in how applicants to the NIST/NRC RAP heard about it.

CHARACTERISTICS OF APPLICANTS AND AWARDEES

From 1965 through February 2007, there were 6,147 applications to the NIST/NRC RAP.[39] From 1965 through February 2007, there were 33,298 applications to the other RAPs, as illustrated in Figure 2-1.[40]

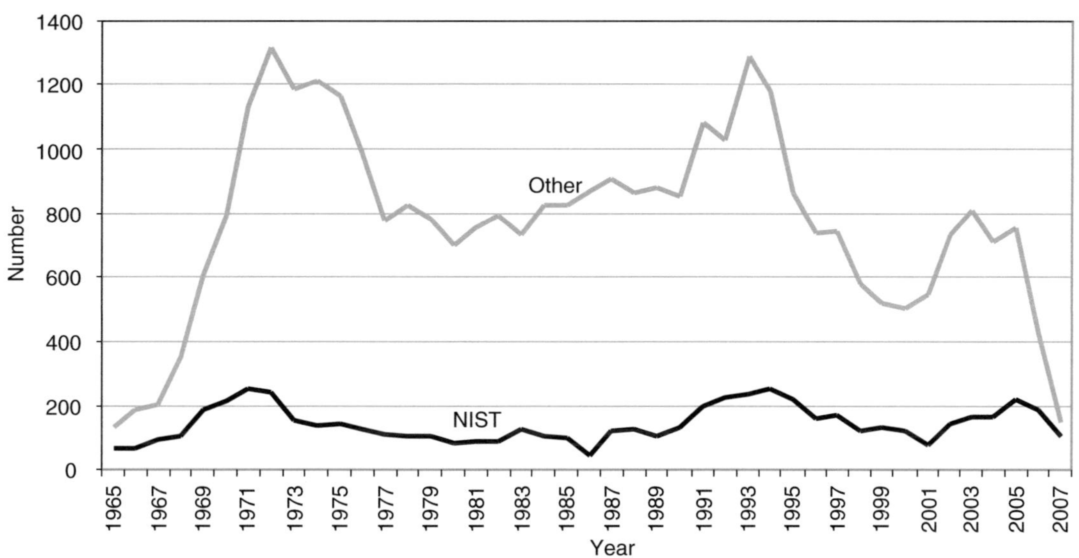

FIGURE 2-1 Number of applications to Research Associateship Programs, by program, 1965-2007.

Note: In 2007, not all application cycles have been completed and recorded in the database.
Source: National Academies, DataRAP Database, tabulations by staff.

Overall, interest in postdoctoral positions has tracked upwards since the 1960s, as noted in the data tables in Appendix E. In general, applications to the NIST/NRC RAP tend to track the other RAPs. Applications peaked in the early 1970s (1970-1972) and again in the mid-1990s (1992-1995). The reduced number of applications between these two time periods likely reflects alternative employment possibilities. In particular, the upswing in the early 1990s may reflect tougher times for finding a regular appointment. Some of the volatility in the NIST/NRC RAP may also be a result of changes in outreach efforts. However, data are not available on the employment preferences of applicants or on efforts to recruit applicants over time. As a final note, in every year, there are more applicants than positions.

The next figure examines the number of applicants to NIST and other Research Associate positions and the number of those accepted to the respective programs. Figure 2-3 compares the acceptance rate for applicants overall to the NIST/NRC RAP and to all other RAPs. (See Appendix E for underlying data.) For those with S&E degrees, from 1965 through February 2007, NIST awardees totaled 1,383—or 22 percent of applicants for NIST/NRC Research Associate positions; while other RAP awardees totaled 9,810—or 29 percent of applicants for

---

[39] There were an additional 35 applications by individuals with Ph.D.s outside S&E, and 58 cases where applicants did not identify their major or the major could not be classified.

[40] There were an additional 797 applications by individuals with Ph.D.s outside S&E, and 1033 cases where applicants did not identify their major or the major could not be classified.

other RAPs. In general, the NIST/NRC RAP is more competitive, in the sense that percentage of awardees to applicants is lower.

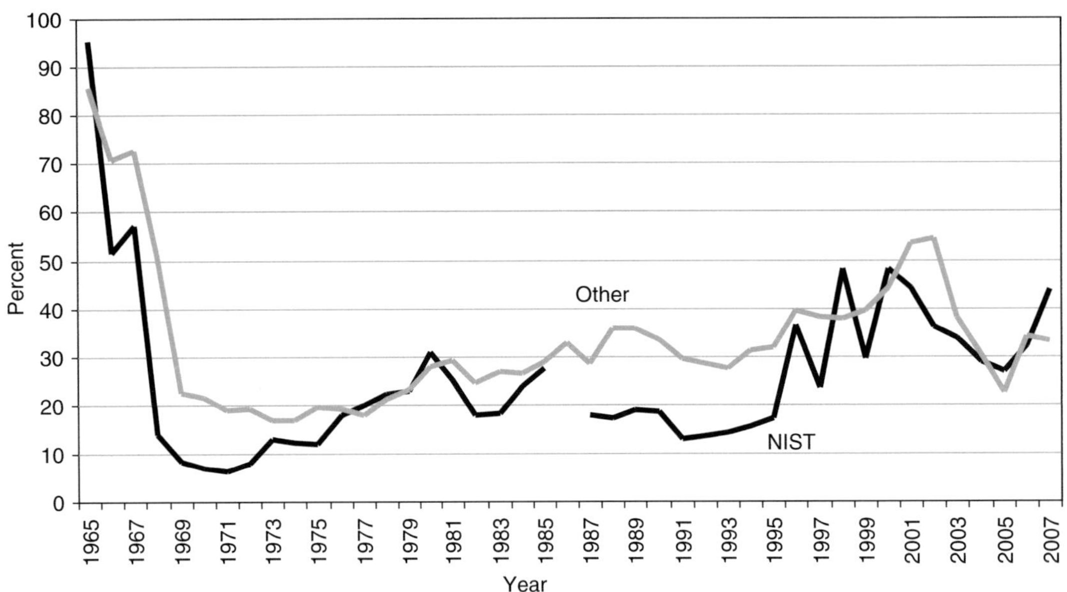

FIGURE 2-2 Percent of awards among applications, by Research Associateship Program, 1965-2007.

Note: No awards were made in the NIST/NRC Program in 1986. In 2007, not all application cycles have been completed and recorded in the database.

Source: National Academies, DataRAP Database, tabulations by staff.

Given that there are many applicants who are not offered a research associateship, due to limitations on the number of positions available, an interesting question is what would happen if the ceiling on research associateships was raised? One way to examine this question is to look at the average scores, assigned by the reviewers, to the applicants who do and do not ultimately get offers. Scores range from 10 to 0, with 10 being the highest, although in practice no one has yet to receive a perfect 10. Reviewers recommended many more candidates than NIST had space for—and some of those who were offered declined. About one-third of all applicants were recommended by the reviewers (but did not receive an award) and about five percent were offered, but declined.

It is likely that the award rates for applicants to the NIST/NRC RAP and the other RAPs vary by a number of characteristics. Next, the applicant pools to the NIST/NRC RAP and the other RAPs are disaggregated by discipline or field, gender, and race/ethnicity.

Discipline

This section presents data on the total pool of potential applicants, applications, and acceptances by discipline. Perhaps reflecting the evolving nature of disciplines over the years, and the fact that applicants are asked to identify a field for each degree received, there are a large number of fields that applicants have identified over the years. There were 1,398 choices applicants could pick. Several are a bit confusing. Applicants picked 536 different fields for their Ph.D.s. To simplify matters, a small group of major categories was created: agricultural sciences and natural resources; biological, biomedical, and health sciences; engineering; mathematical and computer sciences; and the physical sciences to examine. (A list of which fine fields are part of each of these major categories is found in Appendix F. This process was subjective, particularly in terms of whether a field should be included in these major categories or not, but it should not effect the general results presented in the following tables because the least clear fine fields were listed by very few applicants.) See Appendix E for underlying data.

First, what is the proportion of applications by major field? As Table 2-7 illustrates, very few applicants to either the NIST/NRC RAP or other RAPs come from a background in agriculture or natural resources. Applicants to the NIST/NRC RAP are much more likely to come from the physical sciences than applicants to other RAPs, while the reverse is true for applicants who received a Ph.D. in the biological or health sciences.

TABLE 2-9 Applications, by Research Associateship Program and Major Field of Applicants, 1965-2007.

	NIST/NRC RAP		Other RAP	
Field	N	%	N	%
Agric./Nat. res.	2	0.0	377	1.1
Bio/Biomed/Health	80	1.3	6442	19.3
Engineering	1388	22.6	7583	22.8
Math/Comp. sci.	258	4.2	1241	3.7
Physical	4419	71.9	17655	53.0
Total	6147	100.0	33298	100.0

Source: National Academies, DataRAP Database, tabulations by staff.

Second, what is the percentage of awardees, by field? As Table 2-8 shows, the breakdown of awards is fairly consistent with the breakdown of applications. Thus, for NIST postdocs, 72 percent of applications are from doctorates in the physical sciences, and 69 percent of the awards of NIST postdocs are to doctorates in the physical sciences.

TABLE 2-10 Awards, by Research Associateship Program and Major Field of Applicants, 1965-2007.

	NIST/NRC RAP		Other RAP	
Field	N	%	N	%
Agric./Nat. res.	0	0.0	98	1.0
Bio/Biomed/Health	30	2.2	2138	21.8
Engineering	348	25.2	2061	21.0
Math/Comp. sci.	64	4.6	323	3.3
Physical	941	68.0	5190	52.9
Total	1383	100.0	9810	100.0

Source: National Academies, DataRAP Database, tabulations by staff.

Third, what is the ratio of awards to applications, by major field? As Table 2-9 shows, while the NIST/NRC RAP received few applications from doctorates in the biological sciences, broadly defined, and although it selected few life scientists for RA positions, the percentage of doctorates who received Research Associateships from that small group of applicants was relatively quite high—almost 40 percent. Conversely, it was much harder, proportionately, to be awarded a Research Associateship if one's background was in the physical sciences: only 21 percent.

TABLE 2-11 Percent of Awards, by Research Associateship Program and Major Field of Applicants, 1965-2007

Field	NIST/NRC RAP (%)	Other RAP (%)
Agric./Nat. res.	0.0	26.0
Bio/Biomed/Health	37.5	33.2
Engineering	25.1	27.2
Math/Comp. sci.	24.8	26.0
Physical	21.3	29.4
All fields	22.5	29.5

Source: National Academies, DataRAP Database, tabulations by staff.

Gender

A number of recent studies—though this is certainly not a new issue—have focused on efforts to encourage women to pursue careers in S&E. Women are, as noted earlier in the chapter, receiving a growing proportion of Ph.D.s, and yet they have not yet achieved similar results in employment. Some of this is due to the fact that it takes time for women to rise to more senior positions (e.g., full professor in academia), and so more recent gains may not yet be evident in employment characteristics. Nevertheless, it is important that women are receiving encouragement at the beginning of their career—and the NIST/NRC RAP could certainly be a good stepping stone. The following figures examine whether there are differences in the proportion of applicants and awards by gender comparing applicants to the NIST/NRC RAP to other RAPs and to the overall pool.

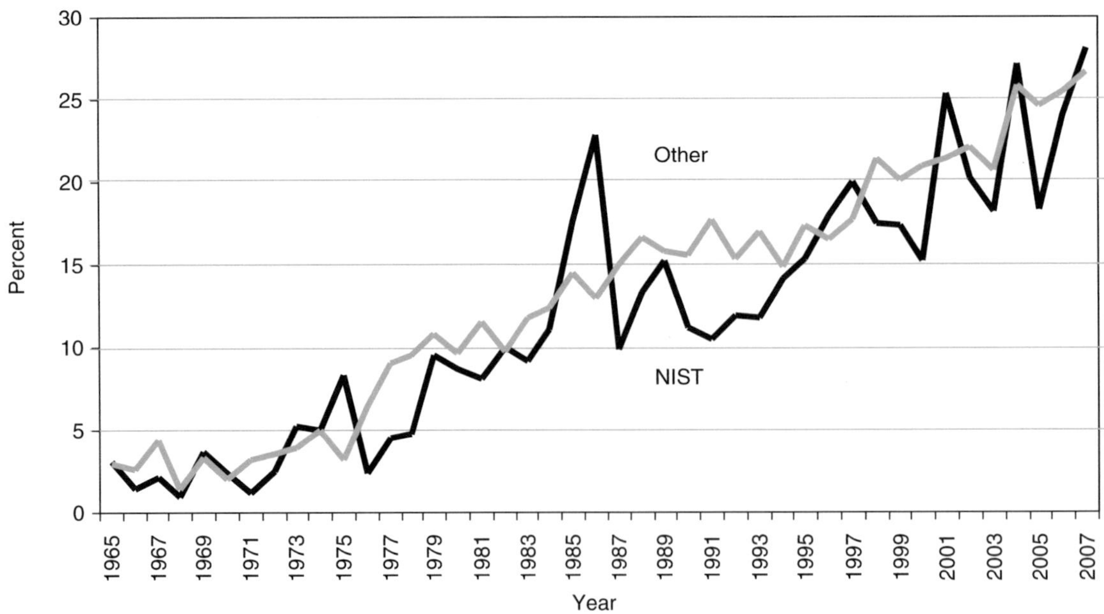

FIGURE 2-3 Percent of applications from women, by Research Associateship Program, 1965-2007.
Source: National Academies, DataRAP Database, tabulations by staff.

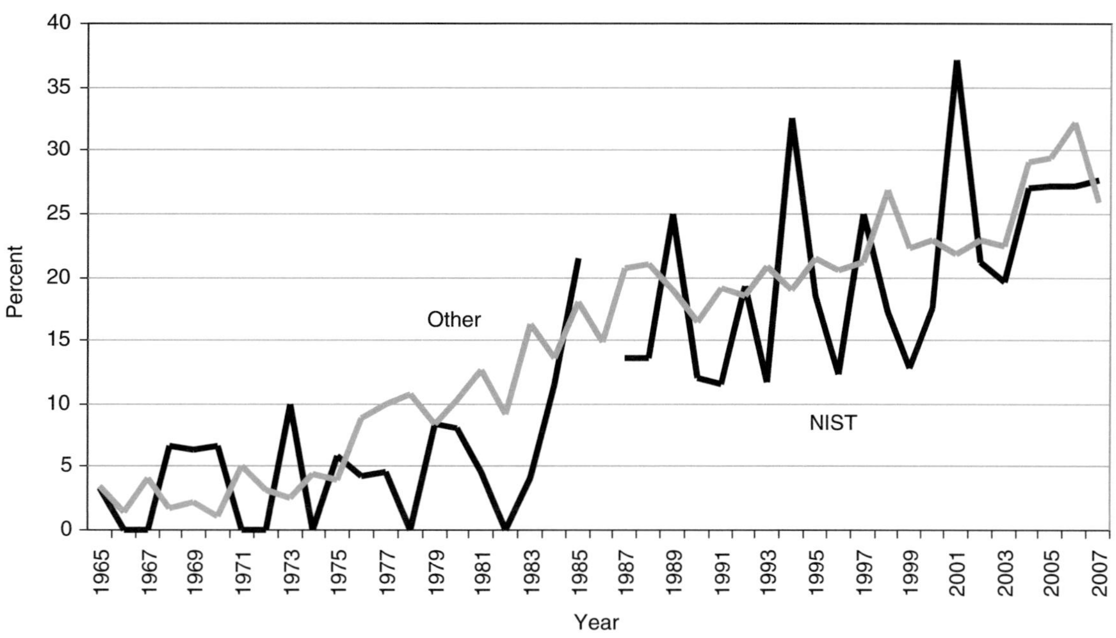

FIGURE 2-4 Percent of awardees who are women, by Research Associateship Program, 1965-2007.

Note: No awards were made in the NIST program in 1986. In 2007, not all application cycles have been completed and recorded in the database.

Source: National Academies, DataRAP Database, tabulations by staff.

Figures 2-3 and 2-4 show that women are increasingly applying to the NIST/NRC RAP; and that the NIST/NRC RAP is as popular as the other RAPs, for applicants. However, the percentage of women applying or receiving an award is lower than the percentage of women receiving Ph.D.s or intending to pursue a postdoc (see Appendix E for comparison data).

Next, the success rate for applications to the two programs are compared, by gender. The question underlying the two following figures is whether applications from men and women, overall, produce similar award rates. For applicants to the NIST/NRC RAP, the two trends are volatile, but roughly similar, as shown in Figure 2-5. As Figure 2-6 illustrates, male and female award rates track quite closely, though applications from women are somewhat more likely to produce awardees.

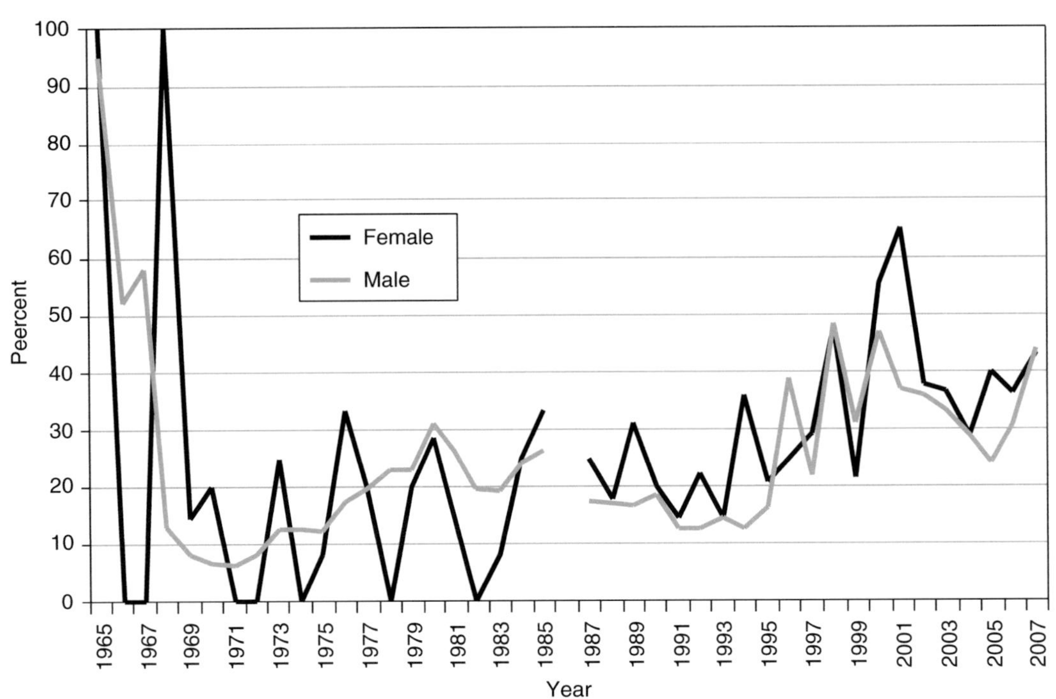

FIGURE 2-5 Success rate of applications to NIST/NRC Research Associateship Program, by gender, 1965-2007.

Note: No awards were made in the NIST program in 1986. In 2007, not all application cycles have been completed and recorded in the database.

Source: National Academies, DataRAP Database, tabulations by staff.

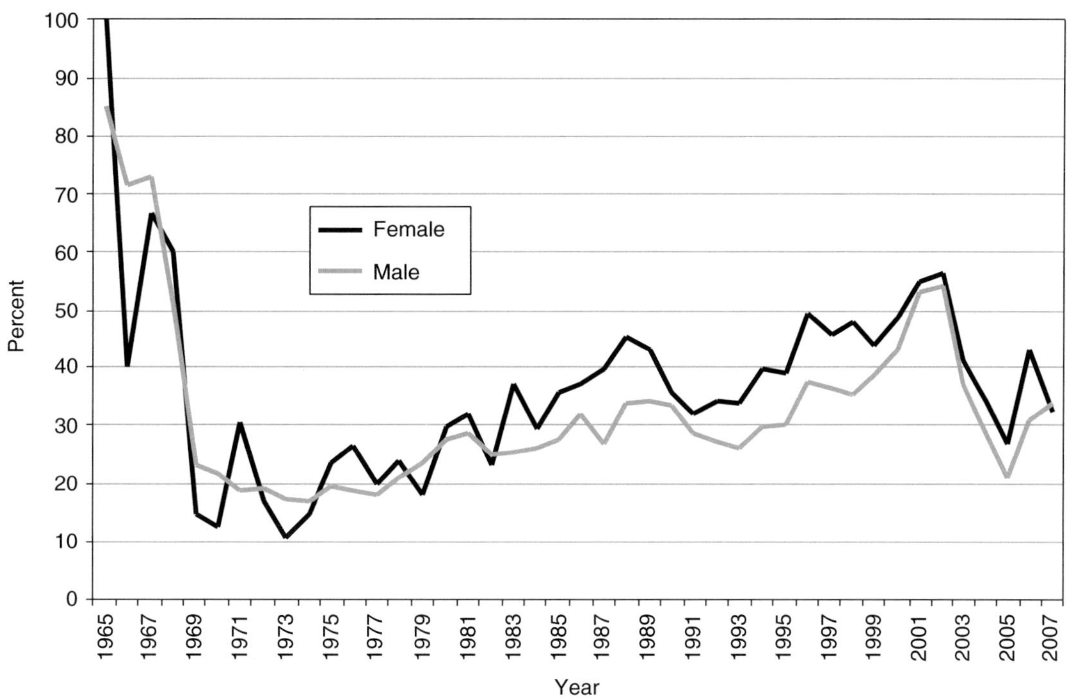

FIGURE 2-6 Success rate of applications to all other Research Associateship Programs, by gender, 1965-2007.
Note: In 2007, not all application cycles have been completed and recorded in the database.
Source: National Academies, DataRAP Database, tabulations by staff.

Race/Ethnicity

This section examines whether there are differences in the proportion of applicants and accepted applicants by race/ethnicity comparing applicants to the NIST/NRC RAP to applicants to the other RAPs and to the overall pool. Two comments preface the analysis. First, because the number of applicants and awardees who are not white is very small (which reflects the relatively smaller number of minorities receiving Ph.D.s in S&E), we grouped the individual race/ethnicity categories together into three groups: white, underrepresented minority, Asian, and unknown. Underrepresented minority here means: American Indian or Alaska Native; Black or African American; Hispanic; or Native Hawaiian or Other Pacific Islander. Asian means Asian or Pacific Islander or Asian. Many people do not fill out this optional question on the application form. Second, these data were not collected prior to 1980. Finally, the percentages here are the proportion of underrepresented minorities among those who indicated a race/ethnicity.

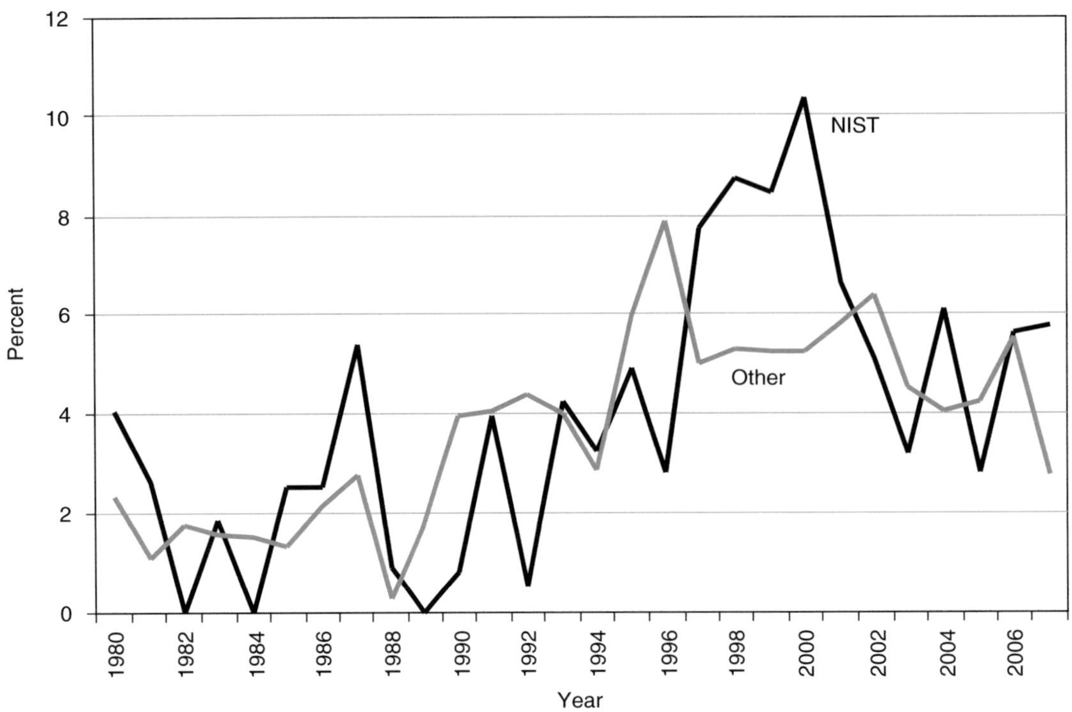

FIGURE 2-7 Percent of applications from underrepresented minorities, by Research Associateship Program, 1980-2007.
Note: In 2007, not all application cycles have been completed and recorded in the database.
Source: National Academies, DataRAP Database, tabulations by staff.

Based on the Figure 2-7 above, NIST has received a greater share of applications from underrepresented minorities over time, although there has been a bit of a drop off in more recently. The decline should be further explored. In general applications from underrepresented minorities are comparable between the NIST/NRC RAP and the other RAPs. However, the percentage of underrepresented minorities applying or receiving an award is generally lower than the percentage of underrepresented minorities receiving Ph.D.s or intending to pursue a postdoc (see Appendix E for comparison data).

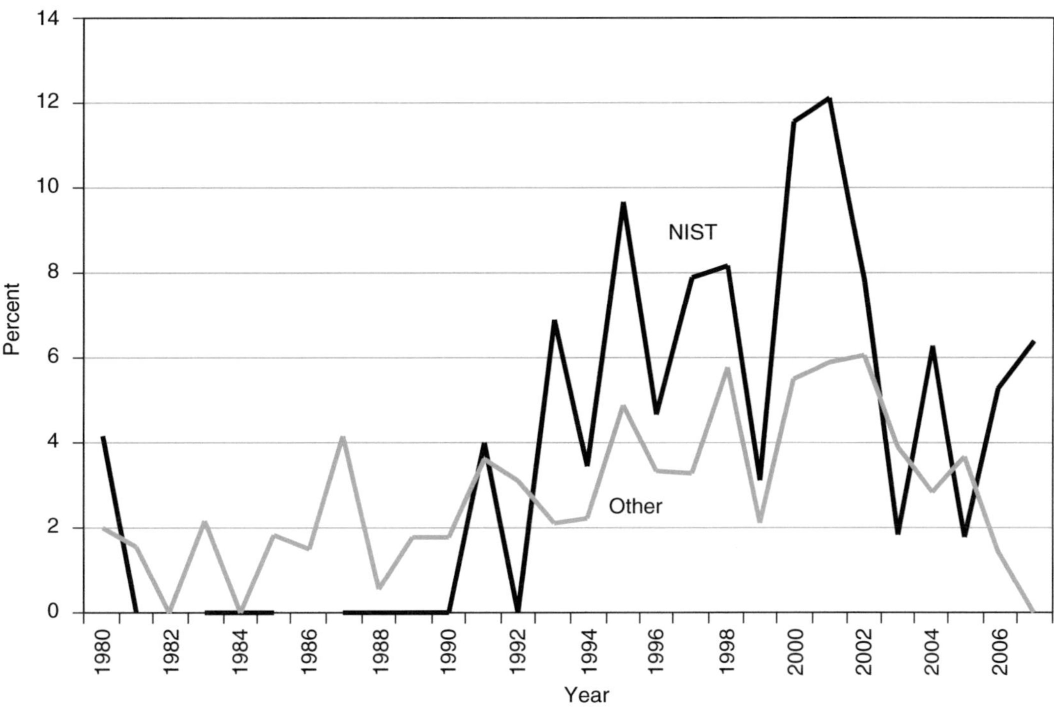

FIGURE 2-8 Percent of awards to underrepresented minorities, by Research Associateship Program, 1980-2007.
Note: No awards were made in the NIST program in 1986. In 2007, not all application cycles have been completed and recorded in the database.
Source: National Academies, DataRAP Database, tabulations by staff.

NIST awards to minorities are also higher in the 1990s than in the 1980s; a positive sign, although not difficult to achieve given that almost no underrepresented minorities received awards in the 1980s. There is a also a decline evident in awards to underrepresented minorities over the past few years, something that should be further scrutinized.

In both figures, the trend line shows more volatility for the NIST/NRC RAP, perhaps in part because one program is being compared to several collectively. However, given data limitations, it appears that the NIST/NRC RAP is often more diverse proportionally in terms of awarding Research Associateships to underrepresented minorities than the other RAPs (taken collectively). In terms of identifying best practices for recruiting applicants, it may be instructive to see if there are different outreach strategies between the NIST/NRC RAP and other RAPs—although in some cases, outreach strategies are conducted by the NRC.

Figures 2-9 and 2-10 repeat the earlier analysis on successful applicants by gender, for race/ethnicity. Here the success rates are compared for underrepresented minorities against whites and Asians. The trend line for underrepresented minorities is quite volatile, as there are few applicants, which makes proportion of awardees among applicants jump around quite a bit. The success rate is similar for the two groups in both the NIST/NRC RAP and other RAPs.

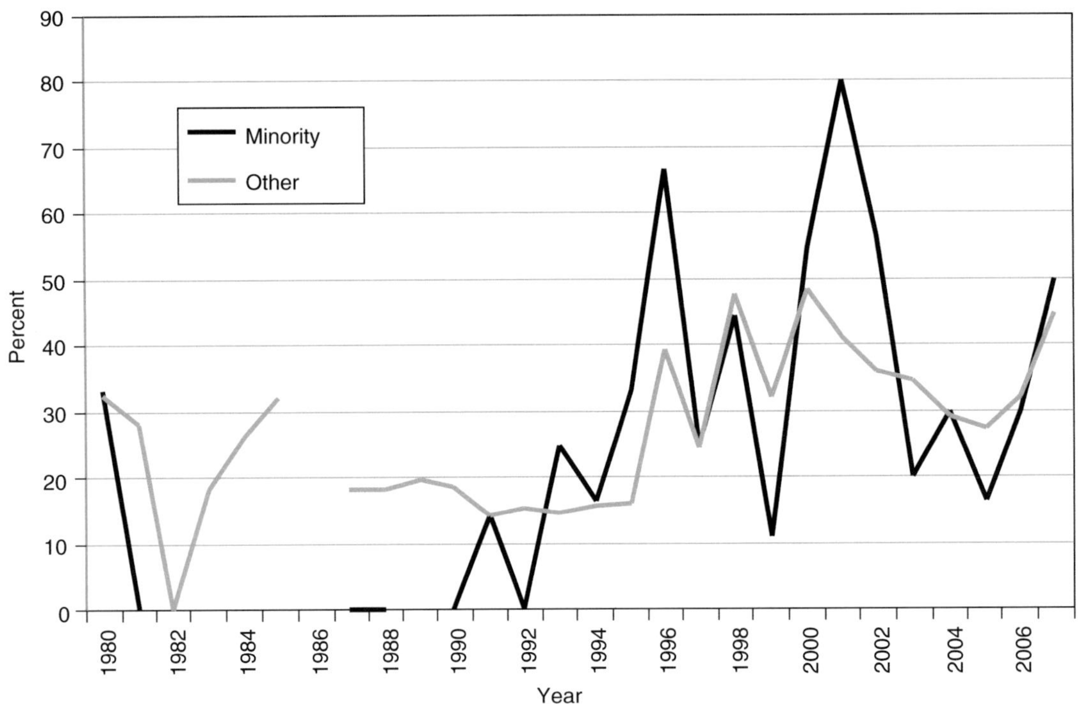

FIGURE 2-9 Success rate of applications to NIST/NRC Research Associateship Program, by race/ethnicity, 1965-2007.

Note: No awards were made in the NIST program in 1986. In 2007, not all application cycles have been completed and recorded in the database.

Source: National Academies, DataRAP Database, tabulations by staff.

FIGURE 2-10 Success rate of applications to all other Research Associateship Programs, by race/ethnicity, 1965-2007.
Note: In 2007, not all application cycles have been completed and recorded in the database.
Source: National Academies, DataRAP Database, tabulations by staff.

Institution

One area of interest for NIST in assessing the NIST/NRC RAP was to examine the doctoral-granting institutions of applicants. Since one goal for NIST was to attract the best and the brightest, there was an expectation that many applicants had received Ph.D.s from what were considered to be top institutions. At the same time, it is true that top candidates can come from a wide range of institutions, so there was also an expectation that there would be some diversity in the range of doctoral-granting institutions of applicants to the program. Applicants were requested on the application form to identify their educational experience and institutions attended. We focused on the institutions that awarded applicants their doctorates, and in particular the top 20 institutions for applications and acceptances are listed below. (Some applicants failed to identify their doctoral-granting institutions: 2 applicants for NIST/NRC RAP applicants and 86 cases among other RAP applicants. These cases were excluded from the analysis. Overall, applications to the NIST/NRC RAP came from Ph.D.s who had received their Ph.D.s from a total of 333 institutions; while applicants to the other RAPs had received doctorates from 1534 institutions. This is likely due in part to the U.S. citizenship requirement for the NIST/NRC RAP. Concerning acceptances, all of those awarded an NIST/NRC RAP identified their doctoral-granting institutions; but 14 of those awarded another Research Associateship did not. NIST/NRC RAs received doctorates from 173 different institutions. RAs for other RAPs received doctorates from 983 different institutions. As Figure 2-11 shows, applicants to NIST come from a large number of institutions.

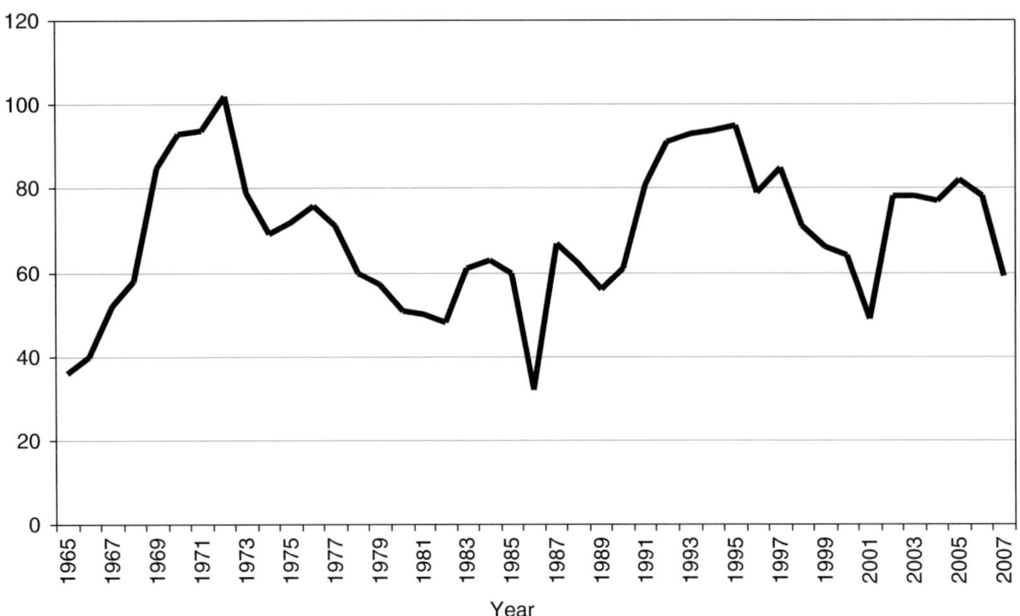

FIGURE 2-11 Number of doctoral-granting institutions for applicants to the NIST/NRC Research Associateship Program, 1965-2007.
Note: In 2007, not all application cycles have been completed and recorded in the database.
Source: National Academies, DataRAP Database, tabulations by staff.

In the next two tables, data are presented on the top 20 doctoral-granting institutions for applicants and awardees for the NIST/NRC RAP and for the other RAPs.

TABLE 2-12 Top 20 Institutions from Which Applications Originated, by Research Associateship Program, 1965-2007

Institutions	NIST/NRC RAP	%	Institutions	Other RAP	%
University of California-Berkeley	222	3.6	University of California-Berkeley	773	2.2
University of Maryland College Park	206	3.4	University of Maryland College Park	587	1.7
University of Colorado	197	3.2	University of Illinois-Urbana-Champaign	572	1.7
Massachusetts Institute of Technology	192	3.1	Massachusetts Institute of Technology	565	1.6
Cornell University	189	3.1	Stanford University	552	1.6
University of Wisconsin-Madison	165	2.7	Cornell University	551	1.6
University of Illinois-Urbana-Champaign	160	2.6	University of Colorado	456	1.3
Harvard University	141	2.3	University of Wisconsin-Madison	455	1.3
University of Michigan	140	2.3	University of California-Los Angeles	435	1.3
Stanford University	134	2.2	University of Michigan	428	1.2
Pennsylvania State University Park	130	2.1	Indian Institute of Science	413	1.2
University of Texas-Austin	107	1.8	Pennsylvania State University Park	396	1.1
Johns Hopkins University	100	1.6	University of Washington	381	1.1
University of Virginia	97	1.6	California Institute of Technology	364	1.1
University of Chicago	96	1.6	University of Arizona	351	1.0
Yale University	92	1.5	University of Texas-Austin	321	0.9
University of California-Santa Barbara	90	1.5	Purdue University	309	0.9
Northwestern University	87	1.4	Johns Hopkins University	302	0.9
Iowa State University	86	1.4	Columbia University	297	0.9
Princeton University	82	1.3	Harvard University	296	0.9
Total	2713	44.3	Total	8804	25.5

Note: In 2007, not all application cycles have been completed and recorded in the database.
Source: National Academies, DataRAP Database, tabulations by staff.

TABLE 2-13 Top 20 Institutions of Awardees, by Research Associateship Program, 1965-2007

Institutions	NIST/NRC RAP	%	Institutions	Other RAP	%
University of Colorado	57	4.1	University of California-Berkeley	251	2.6
University of California-Berkeley	52	3.8	University of Maryland College Park	205	2.1
University of Maryland College Park	52	3.8	Massachusetts Institute of Technology	185	1.9
Massachusetts Institute of Technology	51	3.7	Stanford University	173	1.8
Harvard University	48	3.5	University of Illinois-Urbana-Champaign	158	1.6
University of Illinois-Urbana-Champaign	47	3.4	Cornell University	157	1.6
Cornell University	43	3.1	University of Wisconsin-Madison	145	1.5
Stanford University	40	2.9	Univ of California-Los Angeles	143	1.5
University of Wisconsin-Madison	38	2.7	University of Michigan	142	1.4
University of Michigan	37	2.7	University of Colorado	126	1.3
Pennsylvania State University Park	30	2.2	Pennsylvania State University Park	123	1.3
University of Texas-Austin	29	2.1	California Institute of Technology	122	1.2
Northwestern University	26	1.9	University of Washington	119	1.2
Princeton University	26	1.9	Johns Hopkins University	111	1.1
University of California-Santa Barbara	24	1.7	University of Arizona	106	1.1
Yale University	24	1.7	University of Chicago	97	1.0
Iowa State University	23	1.7	University of California-San Diego	96	1.0
University of Virginia	22	1.6	University of Texas-Austin	96	1.0
California Institute of Technology	20	1.4	University of Florida	94	1.0
University of Washington	20	1.4	University of Minnesota-Twin Cities	93	0.9
Total	709	51.3	Total	2742	27.99

Note: In 2007, not all application cycles have been completed and recorded in the database.
Source: National Academies, DataRAP Database, tabulations by staff.

The tables show quite a bit of overlap. A first point for the NIST/NRC RAP is that the top 20 institutions are all major research institutions. They are all classified as Research I institutions under the 1994 Carnegie classification scheme. Second, 44 percent of all applications to the NIST/NRC RAP come from individuals at 20 institutions and just over half of all awards were made to applicants from 20 institutions.

Next we looked just at NIST, by selected S&E fields. First, in Table 2-14, the number of doctoral-granting institutions, from which applicants and awardees received their Ph.D.s are examined. As Table 2-14 shows, excepting agricultural sciences and natural resources, the applicants come from a large number of institutions. Even in the biological sciences, which produce few applications, there were a relatively high number of universities represented.

TABLE 2-14 Number of Universities from Which Applicants and Awardees Received Their Ph.D.s, by Field, 1965-2007

Discipline	No. of Schools (Applicants)	No. of Schools (Accepted)
Agric. Sciences/Nat. resources	1	0
Bio/Biomed/Health	49	25
Engineering	156	80
Mathematics/CS	97	45
Physical sciences	282	152

Note: In 2007, not all application cycles have been completed and recorded in the database.
Source: National Academies, DataRAP Database, tabulations by staff.

In Tables 2-15 and 2-16, those universities are explored more deeply.

TABLE 2-15 Most Common Doctoral-Granting Institutions of Applicants to the NIST/NRC Research Associateship Program, by Major Field, 1965-2007

Agriculture	Number of Applications from Institution	%
University of California-Berkeley	1	50.0
University of Minnesota-Twin Cities	1	50.0
Total	2	100.0

Bio/Biomed/Health	Number of Applications from Institution	%
Johns Hopkins University	6	7.6
University of Maryland College Park	5	6.3
State University of New York-Stony Brook	4	5.1
University of Virginia	4	5.1
Georgetown University	3	3.8
University of Illinois-Chicago	3	3.8
University of Wisconsin-Madison	3	3.8
University of California-Berkeley	2	2.5
Johns Hopkins University-Medical Insts.	2	2.5
Oregon State University	2	2.5
Purdue University	2	2.5
Rice University	2	2.5
Texas A&M University	2	2.5
U of Maryland School of Medicine	2	2.5
University of Rochester	2	2.5
University of Texas-Austin	2	2.5
Total	46	58.2

Engineering	Number of Applications from Institution	%
Massachusetts Institute of Technology	70	5.0
University of Michigan	54	3.9
University of Illinois-Urbana-Champaign	50	3.6
University of California-Berkeley	46	3.3
University of Colorado	44	3.2
University of Wisconsin-Madison	43	3.1
Northwestern University	43	3.1
Stanford University	43	3.1

Pennsylvania State University Park	40	2.9
Johns Hopkins University	37	2.7
Princeton University	34	2.4
University of Maryland College Park	33	2.4
Carnegie Mellon University	30	2.2
University of Florida	28	2.0
Cornell University	27	1.9
University of Virginia	25	1.8
University of Minnesota-Twin Cities	24	1.7
University of Massachusetts-Amherst	24	1.7
University of Texas-Austin	23	1.7
Total	718	51.7

Mathematics/CS	Number of Applications from Institution	%
University of Wisconsin-Madison	15	5.8
University of Maryland College Park	13	5.0
Northwestern University	11	4.3
Cornell University	11	4.3
Johns Hopkins University	9	3.5
Purdue University	8	3.1
University of California-Berkeley	7	2.7
New York University	7	2.7
University of California-Santa Barbara	6	2.3
State University of New York-Stony Brook	6	2.3
Brown University/RI	6	2.3
Massachusetts Institute of Technology	5	1.9
University of Michigan	5	1.9
University of Colorado	5	1.9
Ohio State University	5	1.9
University of Southern California	5	1.9
Total	124	48.1

Physical sciences	Number of Applications from Institution	%
University of California-Berkeley	172	3.9
Cornell University	155	3.5
University of Colorado	155	3.5
University of Maryland College Park	150	3.4
Harvard University	129	2.9
Massachusetts Institute of Technology	122	2.8
University of Illinois-Urbana-Champaign	109	2.5
University of Wisconsin-Madison	106	2.4
University of Chicago	95	2.2
Stanford University	88	2.0
Pennsylvania State University Park	84	1.9
University of Michigan	81	1.8
University of Texas-Austin	81	1.8
Yale University	71	1.6
University of Virginia	69	1.6
University of California-Santa Barbara	66	1.5

Iowa State University	66	1.5
State University of New York-Stony Brook	56	1.3
California Institute of Technology	55	1.2
University of Florida	55	1.2
Total	1965	44.5

Note: In 2007, not all application cycles have been completed and recorded in the database.
Source: National Academies, DataRAP Database, tabulations by staff.

TABLE 2-16 Most Common Doctoral-Granting Institutions of NIST/NRC Research Associates, by Major Field, 1965-2007

Bio/Biomed/Health	No.	%
Johns Hopkins University	3	10.0
Georgetown University	2	7.0
Johns Hopkins University-Medical Insts.	2	7.0
State University of New York-Stony Brook	2	7.0
Total	9	31.0

Engineering	No.	%
Massachusetts Institute of Technology	21	6.0
University of Michigan	21	6.0
Princeton University	16	4.6
Northwestern University	14	4.0
Pennsylvania State University Park	12	3.4
University of Illinois-Urbana-Champaign	12	3.4
Carnegie Mellon University	11	3.2
Stanford University	11	3.2
University of Massachusetts-Amherst	11	3.2
University of California-Berkeley	10	2.9
University of Colorado	8	2.3
University of Minnesota-Twin Cities	8	2.3
Virginia Polytech Institute and State U	8	2.3
University of Florida	7	2.0
University of Texas-Austin	7	2.0
University of Washington	7	2.0
Iowa State University	7	2.0
Lehigh University	7	2.0
University of Maryland College Park	7	2.0
Johns Hopkins University	6	1.7
Total	211	60.5

Mathematics/CS	No.	%
University of Maryland College Park	4	6.0
University of Wisconsin-Madison	4	6.0
Northwestern University	3	5.0
Pennsylvania State University Park	3	5.0
Cornell University	3	5.0
University of California-Santa Barbara	3	5.0
University of Colorado	2	3.0

University of Washington	2	3.0
Johns Hopkins University	2	3.0
Ohio State University	2	3.0
Syracuse University	2	3.0
Total	30	47.0

Physical sciences	No.	%
University of Colorado	47	5.0
Harvard University	45	4.8
University of Maryland College Park	41	4.4
University of California-Berkeley	40	4.3
Cornell University	35	3.7
University of Illinois-Urbana-Champaign	34	3.6
University of Wisconsin-Madison	31	3.3
Massachusetts Institute of Technology	29	3.1
Stanford University	28	3.0
Yale University	21	2.2
University of Texas-Austin	21	2.2
California Institute of Technology	17	1.8
University of Chicago	17	1.8
University of California-Santa Barbara	16	1.7
University of Virginia	16	1.7
University of Michigan	16	1.7
Iowa State University	16	1.7
Pennsylvania State University Park	15	1.6
Indiana University-Bloomington	13	1.4
Total	498	53.0

Note: In 2007, not all application cycles have been completed and recorded in the database.
Source: National Academies, DataRAP Database, tabulations by staff.

Age

As noted in the start of the chapter, the average age of postdocs has been creeping upwards, in part due to longer time to complete doctorates. Figure 2-12 compares the average age of applicants to and awardees of NIST/NRC RAP and the other RAPs.

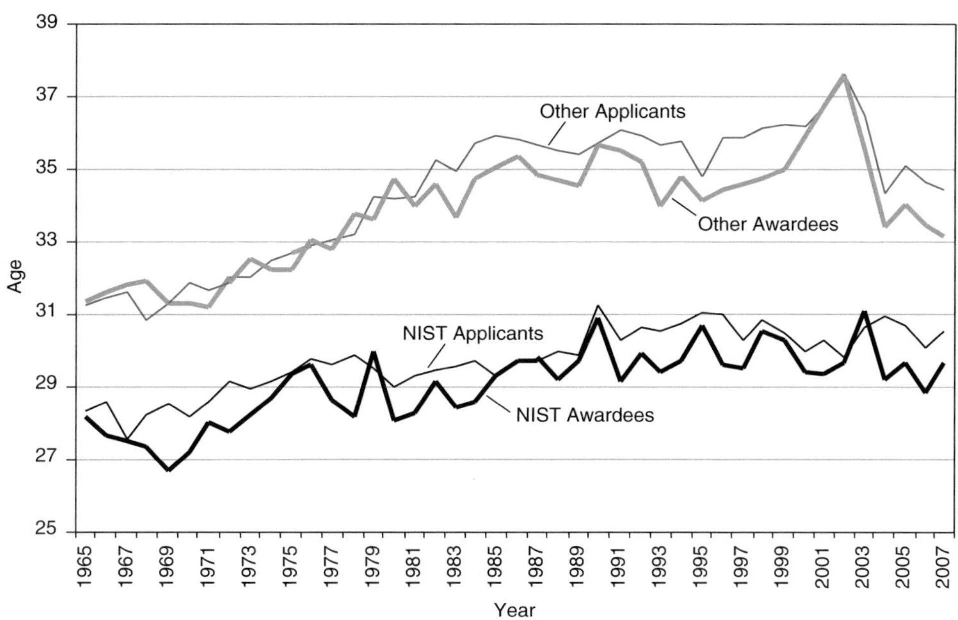

FIGURE 2-12 Average age of applicants and awardees, by Research Associateship Program, 1965-2007.

Note: In 2007, not all application cycles have been completed and recorded in the database.
Source: National Academies, DataRAP Database, tabulations by staff.

As the figure shows, awardees tend to be younger than applicants and applicants to the NIST/NRC RAP and awardees of NIST/NRC Research Associates are younger on average than those who apply for and are awarded other RAPs. It is likely that part of the explanation for this is the group of postdocs coming out of the biological sciences, who are largely absent from the NIST cohort. A second explanation may involve the role of foreign students: NIST RAs are U.S. citizens, who also tend to get to postdoctoral status quicker than international students.

Marital Status

Marital status can be an important demographic characteristic for postdoctoral programs. When many doctorates pursue postdoctoral appointments, they are also at an age when many are married and thinking about starting families. Many scientists are married to other scientists. Knowing this demographic can be helpful in dealing with related issues of: dual-career couples; salary, benefits and cost of living; child care and parental leave. The application form includes a question on marital status. Two categories are available: married and single; although many applicants leave this answer blank. (Additionally, 4 applicants chose "F"—possibly a data coding entry with gender.) Twenty-seven applicants to NIST left this question blank and 368 applicants to non-NIST left it blank. For awards, 2 awardees chose "F," 4 awardees at NIST left it blank, and 121 awardees at other RAPs did not answer the question. The percentages of applicants and awardees that were married or single among those who noted marital status, are examined in Figures 2-13 and 2-14.

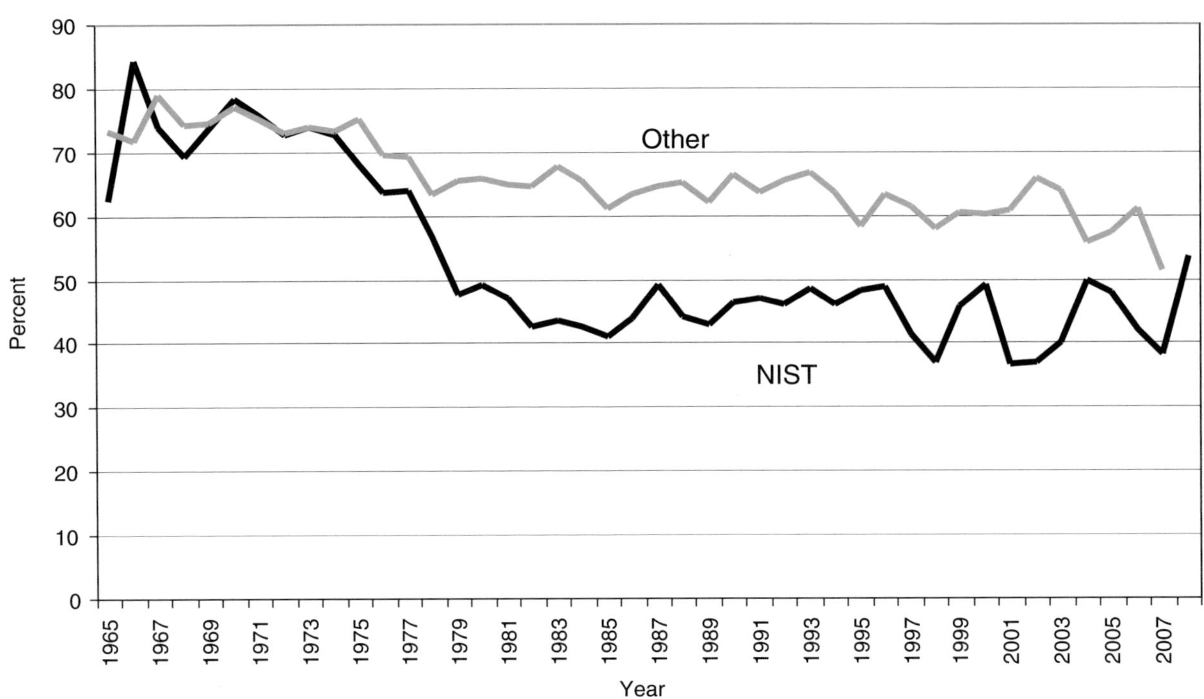

FIGURE 2-13 Percent of applicants who are married, by Research Associateship Program, 1965-2007.

Note: In 2007, not all application cycles have been completed and recorded in the database.

Source: National Academies, DataRAP Database, tabulations by staff.

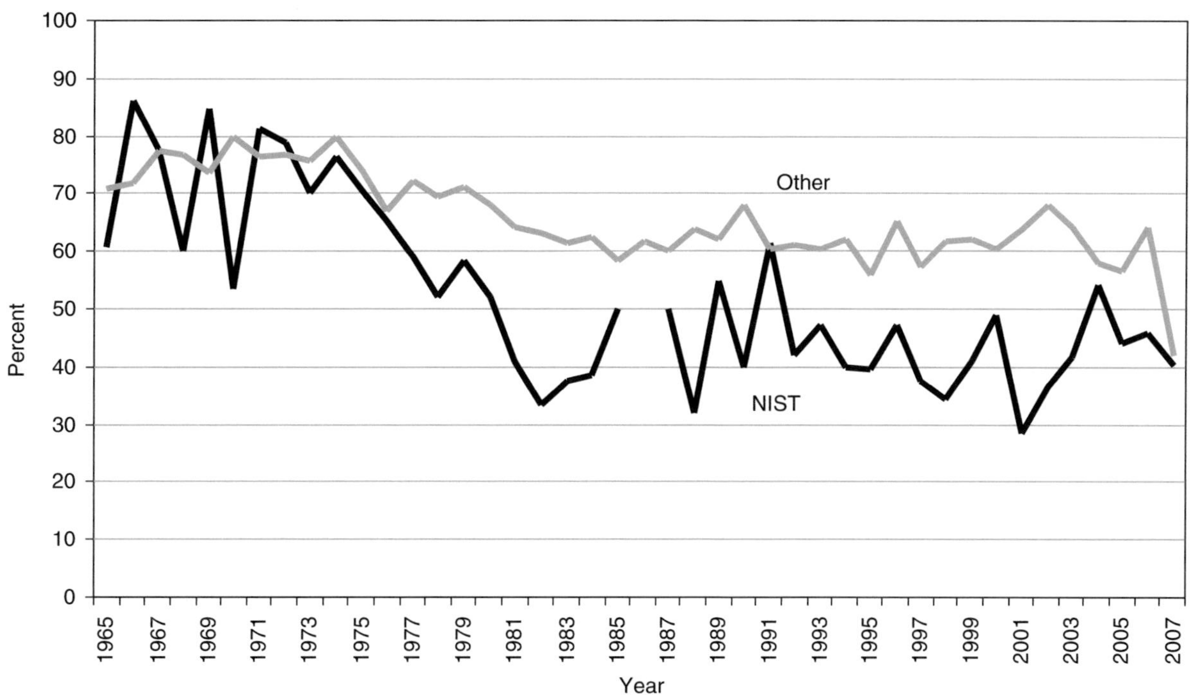

FIGURE 2-14 Percent of awardees who are married, by Research Associateship Program, 1965-2007.

Note: In 2007, not all application cycles have been completed and recorded in the database.
Source: National Academies, DataRAP Database, tabulations by staff.

Applicants and RAs in other RAPs tend to be older and more likely to be married than NIST/NRC RAP applicants and awardees. (See Appendix Table B-10 for the underlying data.) As Figure 2-13 shows, other RAP applicants are more likely to be married. Figure 2-14 shows that other RAP awardees are also more likely to be married. These figures raise a question of whether the NIST/NRC RAP is less attractive to married scientists and engineers or whether some other characteristic of applicants to the NIST/NRC RAP explains the trend that NIST/NRC Research Associates are more likely to be single.

Laboratories

Applicants to the NIST/NRC RAP select a lab on their application form. Over the years NIST has reorganized, which means that some older organizational names are no longer valid, while some recent laboratories may not yet have any applicants. Since 1965, applicants have applied to 18 different parts of NIST. We used the current organizational chart to map older institutional names onto current names (see Appendix E). This was problematic in a number of ways. First, 268 applicants simply put "National Institute of Standards and Technology." Second, several older divisions—e.g., National Engineering Laboratory and the National Measurement Laboratory—map onto multiple contemporary divisions. We combined these situations into a new category: "Multiple," but it can also be thought of as an unknown category. Finally, in spite of the efforts to map the laboratory names, viewing the data over time shows that this is not fully successful. Applicants to Technology Services covered the years 1965 to 1978, but then stopped, although the name still exists, perhaps what the applicants were researching fit better elsewhere. Other labs apparently had no applicants until 1992, again which might reflect an organizational change. Thus, analysis over time, and analysis by race/ethnicity—for which data do not exist prior to 1980—are unwarranted. With available data, two tables can be presented, which focus on applications and acceptances by lab in total, and by gender.

TABLE 2-17 Applications and Awards for the NIST/NRC Research Associateship Program, by Laboratory, 1965-2007

Laboratory	Applications	%	Acceptances	%	Accept. As % of Apps.
Building and Fire Research Laboratory	103	1.7	31	2.3	30.1
Chemical Science and Technology Laboratory	755	12.5	173	12.8	22.9
Electronics and Electrical Engineering Laboratory	355	5.9	101	7.5	28.5
Information Technology Laboratory	250	4.1	43	3.2	17.2
Manufacturing Engineering Laboratory	92	1.5	34	2.5	37.0
Materials Science and Engineering Laboratory	1448	23.9	304	22.5	21.0
Physics Laboratory	674	11.1	203	15.0	30.1
Multiple	2370	39.2	465	34.3	19.6
Total	6047	100.0	1354	100.0	22.4

Note: In 2007, not all application cycles have been completed and recorded in the database.
Source: National Academies, DataRAP Database, tabulations by staff.

The principal finding here is that applications are not evenly distributed among labs. Some labs attract more applicants than others. Likewise, some labs see more research associateships awarded than others. A second finding is that the acceptance rate varies by more than a factor of two—a large range. It may be instructive to determine why this might be the case. Possible answers could focus on the field, other opportunities for recent doctorates in those fields, and outreach by the different labs. Another possible explanation is an intentional "share the wealth" effort.

Table 2-18 continues this examination for gender. Half of the applications from women were to the Chemical Science and Technology Laboratory and the Materials Science and Engineering Laboratory. These are also the two places where a greater proportion of women receive awards. Other labs receive very few applications from women—the Building and Fire Research Laboratory, for example. A second finding is that, in general, the percentages for female applications are similar to the percentage of female awardees. For example, 7 percent of applications to the Electronics and Electrical Engineering Laboratory came from women, while 9 percent of awardees to this lab were women. Additionally, 14 percent of women who applied for the NIST/NRC RAP applied to this lab and 18 percent of women who received awards were in this lab.

Again, it may be instructive to seek out explanations for differences across the labs in terms of the gender (or for that matter, the race/ethnicity) of postdocs. Possible explanations might focus on the role of lab staff in recruiting women candidates; how female-friendly the lab is perceived to be; or the underlying number of female doctorates in fields appropriate to the research of each lab.

TABLE 2-18 Applications and Awards for the NIST/NRC Research Associateship Program, by Laboratory and Gender, 1965-2007

Laboratory	Female Applications (N)	Female Applications (%)	Female of Total Applications (%)	Female Awards (N)	Female Awards (%)	Female of Total Awards (%)
Building and fire research laboratory	5	0.7	4.9	1	0.5	3.2
Chemical science and technology laboratory	168	23.6	22.3	54	26.7	31.2
Electronics and electrical engineering laboratory	51	7.2	14.4	18	8.9	17.8
Information technology laboratory	28	3.9	11.2	3	1.5	7.0
Manufacturing engineering laboratory	10	1.4	10.9	1	0.5	2.9
Materials science and engineering laboratory	186	26.1	12.8	56	27.7	18.4
Multiple	176	24.7	7.4	31	15.3	6.7
Physics laboratory	89	12.5	13.2	38	18.8	18.7
Total	713	100.0	11.8	202	100.0	14.9

Note: In 2007, not all application cycles have been completed and recorded in the database.
Source: National Academies, DataRAP Database, tabulations by staff.

Declined Offers

One concern heard at focus groups was that NIST was not quick enough at offering applicants awards. Individuals were accepting other positions instead of coming to NIST. This section looks at offers that were accepted and those that were declined. Table 2-19 shows the number and percentage of offers declined among those who accepted or declined. On average, 7 people per year decline to accept an award from NIST, compared with an average of 68 people per year for other RAPs.

TABLE 2-19 Number of Individuals Offered a Research Associateship Who Decline, by Research Associateship Program, 1965-2007

	NIST/NRC RAP		Other RAPs	
Year	Number	%	Number	%
1965	0	0.0	4	3.4
1966	1	2.7	10	6.9
1967	0	0.0	8	5.2
1968	9	37.5	36	16.7
1969	8	33.3	75	35.4
1970	14	48.3	76	30.6
1971	7	30.4	89	29.2
1972	19	50.0	107	29.7
1973	9	31.0	103	33.8
1974	13	43.3	101	33.0
1975	6	26.1	88	27.8
1976	6	20.7	92	32.4
1977	12	35.3	68	32.7
1978	16	41.0	103	36.9
1979	20	45.5	85	32.1
1980	11	30.6	114	36.9
1981	23	51.1	87	28.2
1982	3	15.8	70	26.4
1983	11	31.4	79	28.7
1984	14	35.0	104	32.1
1985	14	33.3	138	36.6
1986	N/A	N/A	71	20.0
1987	8	26.7	65	20.0
1988	14	38.9	71	18.6
1989	5	20.0	54	14.6
1990	5	16.7	68	19.2
1991	2	7.1	66	17.2
1992	8	20.5	64	18.0
1993	5	12.8	82	18.8
1994	11	21.6	66	15.2
1995	2	5.0	66	19.4
1996	1	1.7	55	15.8
1997	6	13.0	53	15.8
1998	1	1.7	40	15.4
1999	2	4.9	31	13.0

2000	6	9.5	46	17.2
2001	0	0.0	41	12.3
2002	1	1.9	59	12.9
2003	0	0.0	46	13.0
2004	6	11.1	44	16.9
2005	15	20.3	33	16.3
2006	9	13.2	30	17.0
2007	1	2.1	11	18.0
Total	324	19.0	2799	22.2

Note: No awards were made in the NIST program in 1986. In 2007, not all application cycles have been completed and recorded in the database.
Source: National Academies, DataRAP Database, tabulations by staff.

Since the number of research associates is relatively small for NIST, these declinations may be more noticeable. Interestingly, the rate of declined offers is itself declining. Perhaps the program is becoming more prestigious, the job market has changed, or the offer process has changed.

PRELIMINARY RESULTS

Outreach efforts produce more qualified applicants than NIST has slots to fill for research associates; and the pool of applicants includes many from top research institutions and is increasingly diverse. Overall, 22 percent of applicants were awarded an appointment—a lower ratio than for RAPs elsewhere. Women are increasingly applying to the NIST/NRC RAP and being awarded research associateships. The NIST/NRC RAP seems to be as popular as the other RAPs for women. Underrepresented minorities are also increasingly applying to the NIST/NRC RAP and being awarded research associateships. For applicants to the NIST/NRC RAP and awardees, at least half came from 20 of the top doctoral-granting institutions in the United States

Applicants and awardees to the NIST/NRC RAP differ from their counterparts in the other RAPs. Since 1990, underrepresented minorities are proportionately more likely to be awarded a NIST/NRC Research Associateship than a research associateship in another program. Applicants to, and awardees of, NIST research associateships are younger on average than those who apply for other research associateships. NIST/NRC RAP applicants and awardees are more likely to be single. They are more likely to have Ph.D.s in the physical sciences than biological. The majority of awards go to doctorates from the physical sciences. But, because there are so many applications from this discipline, only about one in five applicants with this background receive awards.

Preliminary analysis suggests that labs receive different amounts of applications and awards are not made uniformly across different labs. Some awardees do decline NIST/NRC Research Associateships, though the percentage of declined offers is often lower than that for the other RAPs and has declined over time.

RECOMMENDATIONS

1. **NIST should conduct an evaluation of outreach efforts.**
    a. To conduct such an evaluation, data need to be collected. In this regard, the question on the application about how applicants hear about the program is

helpful and should be retained. However, the "Other" category should be further analyzed and a choice of "Website" should be added as a category.

b. Additional data could be collected from NIST personnel and former or current NIST RAs. Such data could be used to answer such questions as:
    i. What mechanisms do NIST personnel and RAs use to interact with potential applicants and
    ii. Which mechanisms seem to work best?
    iii. Has there been any effort to focus specifically on diversity? How?

   Such research could be undertaken via a combination of expert panels or surveys of NIST staff and current or former RAs to answer the first and third questions and to provide information for an assessment of the second question. Information should also be collected on the costs for individual outreach efforts (e.g., money spent on advertisements, time spent meeting with graduates) to compare to the benefits (how many applicants come from each individual outreach type).

c. A second step to facilitate an evaluation of outreach efforts is to identify metrics for quantifying value obtained from different outreach strategies, such as hits to the website or number of graduate students met with at professional meetings.

d. Examine individual outreach strategies for return on investment. This could include such strategies as assessing the NIST website for usability and informational content or assessing the return on advertising in publications. As part of the assessment of the NIST Web site, NIST could consider adding contact information for research advisors to facilitate a dialogue between potential applicants and relevant NIST staff.

e. Finally, consider whether there might be other outreach strategies that are being underused currently, and which might have potential value, such as direct mail to deans, department heads and other university administrators.

f. In addition, it is important to determine if any groups of graduate students—and potential applicants—who would make good candidates for the NIST/NRC RAP are unaware of the Program and how one applies. It would be difficult to craft a random sample of graduate students, but a limited survey might be possible.

2. **NIST should conduct an evaluation of individuals who decline offers of Research Associateships.** This could be done as a telephone interview or via a survey. As there are only a few people who decline each cycle, the burden would be relatively small. Two basic questions should be asked of those who are awarded but decline: (1) why are you declining, and (2) what are you planning to do instead?
3. **The NRC should amend the application form.** The number of fields should be reduced, in particular by collapsing very similar labels and by removing labels that are for multiple fields (e.g., "Biophysics Physics Biochemistry"). At least with regard to Ph.D. fields, an example of a smaller field list is found in the NSF's Survey of Earned Doctorates (see Appendix B).
4. **The NRC should update the DataRAP database to replace organizational names (e.g., institutes or labs) that no longer exist at NIST with current equivalents.**

3

Research Associates' Experiences

Research Associates (RAs) spend up to two years in residence at NIST conducting research. During the expert panels we asked current and former research associates several questions about their experiences at NIST. Among the current RAs, they were quick to point out that this appointment was their first professional activity after graduate school and their first experience outside academia. Current research associates reported that one could have a fair amount of research freedom at NIST, but that it varied across labs and groups. Variation across different parts of NIST was perhaps one of the most important themes that emerged from the expert panels.

The current research associates felt that the position gave them a broader view of science. They were overwhelmingly satisfied with the experience, noting that NIST was a great place to work. They also noted that being in the NIST/NRC RAP gave them better benefits and more access on the NIST campus. Finally, some participants felt that access to resources at places such as NIST were likely to be better than in other settings, such as academia.

On the negative side, they felt that they faced additional bureaucracy by being in the program. Other concerns of current research associates included that there were no teaching opportunities and that there were few graduate students in their labs. Current RAs seemed to refer to the NIST/NRC RAP in terms of their recent academic experience and while some of them reported that their colleagues treated them as equals, others felt like they were the "lowest on the totem pole." Internal review of manuscripts was also seen as something of a burden, again perhaps in reference to graduate school, although participants noted that in practice the delay caused by internal review was not long and that having additional reviewers helped improve their work.

Current research associates' suggestions for improving the program were to increase the salary, possibly increase the duration of the appointment, and create a better family leave policy. In referencing the salary, research associates believed the salary to be higher than many other—especially academic—postdoctoral positions. The annual base salary for a NIST RA is $60,000. The Sigma Xi postdoctoral survey noted that the average salary for postdocs was $38,000 (Davis, 2005). A recent survey of postdocs in the life sciences found that in 2006 the average salary for postdocs was $52,750 in industry; $50,000 in government; $40,000 in medical schools; and $38,000 in academia (Austin, 2006). They did, however, identify at least one postdoctoral position that they thought had a higher salary (Sandia), but primarily their motivation for increasing the salary tended to do with the high cost of living in the Washington metro area, where the majority of NIST RAs work. Participants mentioned that one reason for considering longer postdoctoral terms had to do with research associates' shifting fields of research. They noted that moving into a new research area required significant time to get up to speed in that area and then at the end, there was a rush to conduct research and publish findings. Participants also believed that family leave was insufficient. They were concerned that if they took time off because of a new baby, they would simply lose time off their appointment, which was not replaced.

Former research associates (and as noted in Chapter 1, these were limited to those who stayed at NIST) also held the program in high esteem. Perhaps more so than the current research associates, they viewed the program as very prestigious. The postdoc was seen as a good stepping stone, particularly for preparing for a government career. Former RAs saw several benefits to the program, including: collaboration (including the range of scientists or engineers worked with), a good stepping stone on the career pathway, and a good way to get a job at NIST. Former RAs still working at NIST were more likely to view their colleagues as treating them as professional colleagues, although this of course might be expected given that the former RAs were currently on temporary or career conditional appointments.

On the negative side, some participants noted that the postdoc was not as helpful for RAs seeking academic positions in liberal arts colleges. Former RAs did note that NIST was not, in their view, set up to provide a lot of mentoring and that advising and mentoring were less common at NIST. They felt that RAs who needed a lot of mentoring or hand-holding did not do well at NIST, but that self-sufficiency and independence were characteristics of more successful (perhaps more satisfied) RAs. Neither described as a negative nor a positive, it was noted by former RAs that some RAs do not end up researching what they originally proposed in their applications.

Former RAs made a few suggestions regarding possible ways to improve the program. Like the current RAs, they saw the salary as a bit low, again in reference to the high cost of living they felt in the Washington metro area. An interesting discussion took place over whether RAs could get raises. They found information on raises to be less transparent then they preferred and some former RAs believed that because of bureaucratic maneuvering they could not get raises. Again, there were different opinions based largely on which lab the participants were familiar with. Former RAs felt that the two-year time period was appropriate for the appointment. Finally, former RAs were unsure of who in NIST was the "champion" for RAs (in reference to who could help them if they experienced any problems).

Advisors and division leaders offered a different view of the benefits of the program, focusing more on the benefits to NIST. They noted the program was a good way to recruit potential employees, try people out for two years, and retain good people. Some participants noted that the program was the "primary" way NIST recruited. They also see many benefits in having the RAs at NIST, including: covering a wider range of expertise, getting research done, bringing in new ideas, doing innovative research, and helping NIST connect to universities.

Advisors and division leaders also offered different suggestions for improving the program. They felt that better recruiting was needed. Participants believed that personal relationships were key to recruiting. Second, they felt that there should be a NIST-wide support mechanism for RAs. They wanted to see more activities for RAs to interact and network. Third, like the former RAs, they felt that two years was an appropriate duration, although they were willing to explore a third year option for selected RAs. Advisors and division leaders felt that in some areas there were not enough applicants. One way to get more applications was to consider opening up the award to non-citizens, although they note that non-citizens face more restrictions in working at NIST. An alternative idea mentioned was to open the award up to green card holders. Overall, participants in the three expert panels felt that the program had myriad benefits to the individual RAs and to NIST.

In terms of quantifying those benefits, there is less to work with. The two principal sources are an evaluation form filled out by the RAs when they complete their tenure, or term, at NIST; and an evaluation of the RAs by their research advisor. (See Appendixes 8 and 9 for the forms.) Neither form was viewed by many as required and was not filled out by most RAs or advisors. Only 253 NIST/NRC RAs partially or completely filled out the final report for a response rate of about 20 percent.[41] (Additionally, in the beginning of the program very few RAs filled out the form compared with the last t10 years or so.) For RAs in the other RAPs, 6,936 partially or completely filled out the form, for a response rate of about 69 percent.[42] The research advisor's evaluation form was not filled out by a sufficient number of advisors at NIST (less than 20 did so) to include information from the form in an assessment of the RAs. Therefore, results drawing on these data should be taken cautiously. There may be nonresponse bias, particularly if RAs who had better experiences or more positive outcomes also tended to be more likely to fill out the form. Additionally, in some cases, the forms ask for historical information, such as the number of presentations RAs gave. It is possible that some respondents answered inaccurately, either unintentionally or intentionally. But this seems unlikely, as respondents are asked to provide details (e.g., title, publication) rather than counts (i.e., the number of journal articles published). In any case, there was no way in this study to independently confirm these data, as respondents were not identified by name. However, future analysis, if confidentiality concerns could be met, could undertake a CV analysis of a sample of respondents to compare with responses on the final reports.

Hopefully, as the forms are completed more in the future, they will provide more data for NIST. There is no evaluation by NIST employees of the value of the program overall, save for a question on the research advisor's evaluation form. As an open-ended question, it is difficult to quantify answers.

The final report completed by RAs provides information in two areas: productivity of the research associate during the tenure and their views of the Program. Each of these areas is addressed in turn.

PRODUCTIVITY DURING THE POSTDOCTORAL APPOINTMENT

Productivity measures focus on both quantity and quality metrics, although only the quantitative ones are entered into the NRC's DataRAP database. Measures focus on publications (peer-reviewed journals); books, book chapters, other publications; patents, international or domestic presentations, seminars or lectures delivered, manuscripts in preparation and manuscripts submitted; and professional awards received. Selected outcome measures are examined.

Journals

In the final report, RAs are asked to provide complete citations for their publications, including journal articles in peer-reviewed journals. The data which is entered into the database consists of counts of journal articles. Fifty-five percent (of 253) NIST/NRC RAs and 34 percent (of 6,936) of RAs from other RAPs provided information on journals. As Table 3-1 shows, for those who responded, NIST/NRC RAs published between 0 and 13 articles; while RAs from other RAPs published between 0 and 36 articles during their appointments. On average, NIST/NRC RAs published slightly fewer articles in peer-reviewed journals than RAs in other RAPs (2.3 to 2.7). This difference is statistically significant at the 0.05 level.[43] However, a greater percentage of responding NIST/NRC RAs published at least one article.

[41] Counting partial interviews as respondents.

[42] Counting partial interviews as respondents.

[43] Based on an unpaired t-test with unequal variances. Note the different sample sizes and that the samples are not normally distributed.

TABLE 3-1 Number of Articles Published in Peer-Reviewed Journals by Research Associates, by Program

No. of Articles Published	NIST/NRC Research Associates		Other Research Associates	
	No.	%	No.	%
0	22	15.9	553	23.5
1	40	29.0	506	21.5
2	23	16.7	369	15.7
3	21	15.2	296	12.6
4	11	8.0	174	7.4
5	10	7.2	125	5.3
6	4	2.9	89	3.8
7	4	2.9	72	3.1
8	2	1.4	64	2.7
9		0.0	31	1.3
10		0.0	25	1.1
11		0.0	12	0.5
12		0.0	8	0.3
13	1	0.7	7	0.3
14		0.0	3	0.1
15		0.0	3	0.1
16		0.0	2	0.1
17		0.0	3	0.1
18		0.0	1	0.0
19		0.0	2	0.1
21		0.0	2	0.1
23		0.0	2	0.1
25		0.0	1	0.0
30		0.0	1	0.0
36		0.0	1	0.0
N	138	100.0	2352	100.0

Source: National Academies, DataRAP Database, tabulations by staff.

Presentations

Research associates were also asked to provide details on each of their presentations (both domestic and international) made during their appointment. Fifty-six percent (of 253) NIST/NRC RAs and 35 percent (of 6,936) of RAs from other RAPs provided information on domestic presentations. NIST/NRC RAs gave between 0 and 14 domestic presentations, while RAs from other RAPs gave between 0 and 27 presentations, although more than 90 percent of all RAs gave at least one presentation. As Table 3-2 shows, for those who responded, on average, NIST/NRC RAs gave slightly more domestic presentations at scientific meetings or conferences than RAs in other RAPs did (3.5 to 3.0). This difference is statistically significant at the 0.05 level.[44] And NIST/NRC RAs were more likely to give at least one domestic presentation than their counterparts at other RAPs.

[44] Based on an unpaired t-test with unequal variances. Note the different sample sizes.

TABLE 3-2 Number of Domestic Presentations, by Program

No. of Domestic Presentations Made	NIST/NRC Research Associates		Other Research Associates	
	No.	%	No.	%
0	9	6.4	398	16.5
1	23	16.3	455	18.9
2	32	22.7	389	16.1
3	12	8.5	373	15.5
4	24	17.0	238	9.9
5	16	11.3	179	7.4
6	10	7.1	118	4.9
7	8	5.7	88	3.7
8	3	2.1	42	1.7
9		0.0	38	1.6
10	1	0.7	31	1.3
11	1	0.7	12	0.5
12		0.0	18	0.7
13	1	0.7	10	0.4
14	1	0.7	4	0.2
15		0.0	4	0.2
16		0.0	5	0.2
17		0.0	4	0.2
18		0.0	2	0.1
19		0.0	1	0.0
27		0.0	1	0.0
N	141	100.0	2410	100.0

Source: National Academies, DataRAP Database, tabulations by staff.

For international presentations, 49 percent (of 253) NIST/NRC RAs and 31 percent (of 6,936) of RAs from other RAPs provided information. NIST/NRC RAs gave between 0 and 7 international presentations, while RAs from other RAPs gave between 0 and 25 presentations. Among those who responded, a greater percentage of NIST/NRC RAs gave at least one international presentation. As Table 3-2 shows, neither NIST/NRC RAs nor RAs in other RAPs give many international presentations, however, with an average of one presentation. While the NIST/NRC RAs have a slightly lower average than RAs in other RAPs (0.9 to 1), the difference is not statistically significant.

TABLE 3-3 Number of International Presentations, by Program

No. of International Presentations	NIST/NRC Research Associates		Other Research Associates	
	No.	%	No.	%
0	64	51.2	1190	54.9
1	35	28.0	480	22.1
2	14	11.2	247	11.4
3	6	4.8	107	4.9
4	3	2.4	52	2.4
5	1	0.8	40	1.8
6	1	0.8	17	0.8
7	1	0.8	8	0.4
8		0.0	4	0.2
9		0.0	7	0.3
10		0.0	4	0.2
11		0.0	1	0.0
12		0.0	7	0.3
13		0.0	4	0.2
25		0.0	1	0.0
N	125	100.0	2169	100.0

Source: National Academies, DataRAP Database, tabulations by staff.

Patents

Patents during a postdoctoral appointment were quite rare for Research Associates. Fifty-seven percent (of 253) NIST/NRC Research Associates and 28 percent (of 6,936) of research associates from other RAPs provided information on patents received. Among those who responded, as presented in Table 3-4, similar percentages of NIST/NRC Research Associates and Research Associates in other RAPs indicated that they had one or more. Numerically, both NIST/NRC Research Associates and Research Associates of other RAPs reported an average of 0.1 patents, with no statistically significant difference.

TABLE 3-4 Number of Patents, by Program

No. of Patents	NIST/NRC Research Associates		Other Research Associates	
	No.	%	No.	%
0	112	93.3	1733	90.6
1	5	4.2	127	6.6
2	2	1.7	31	1.6
3	1	0.8	16	0.8
4		0.0	3	0.2
5		0.0	2	0.1
N	120	100.0	1912	100.0

Source: National Academies, DataRAP Database, tabulations by staff.

Awards

Awards were also rare for Research Associates in any Program. Twenty-nine percent (of 253) NIST/NRC Research Associates and 10 percent (of 6,936) of research associates from other RAPs provided information on awards received. As with patents the results were quite similar, both in percentage terms and numerically. Among those who responded, as Table 3-5 illustrates, 12 percent of NIST/NRC Research Associates had received an award, compared with 16 percent of other Research Associates who had received one or more awards. Numerically, on average, NIST/NRC Research Associates received fewer of these rare awards than did Research Associates of other RAPs (0.1 to 0.2). This difference is statistically significant at the 0.05 level.[45] However, quantitative measures of awards are not very helpful, as some awards are clearly more important than others.

TABLE 3-5 Number of Awards, by Program

	NIST/NRC RAP		Other RAPs	
No. of Awards	No.	%	No.	%
0	64	87.7	557	84.0
1	9	12.0	94	14.0
2		0.0	10	2.0
3		0.0	2	0.0
4		0.0	2	0.0
8		0.0	1	0.0
N	73	100.0	666	100.0

Source: National Academies, DataRAP Database, tabulations by staff.

RESEARCH ASSOCIATES' VIEWS OF THE PROGRAM

The evaluation form asks Research Associates to rate the RAP on six dimensions on a scale of 1 to 10, with 1 representing poor and 10 meaning excellent. Not surprisingly, given the positive feedback heard in the expert panels, most respondents had a very positive view of the program. Note that the number of responses to this part of the questionnaire was much lower than other parts, possibly because it comes at the end of the questionnaire. Here the response rate is around 15 to 20 percent. As a result all of these findings need to be taken cautiously.

Short Term

This dimension asks Research Associates to evaluate the program in terms of the development of knowledge, skills, and research productivity. Thirty-seven percent (of 253) NIST/NRC Research Associates and 14 percent (of 6,936) of research associates from other RAPs provided information on this category. Among those who responded, as seen in Table 3-6, the average response was an 8.7, with NIST/NRC Research Associates answering 8.5 on average

[45] Based on an unpaired t-test with unequal variances. Note the different sample sizes and that the samples are not normally distributed.

and Research Associates of other RAPs answering 8.7. The difference between the two groups was not statistically significant.

TABLE 3-6 Research Associates' Appraisal of the Short-Term Value of the Research Associateship Program, by Program

On A Scale of 1-10 (Poor-Excellent), Please Rate the Following: Short Term Value	NIST/NRC RAP	Other RAPs
1	1	9
2		2
3	1	4
4		6
5	1	31
6	4	21
7	8	81
8	25	217
9	28	202
10	26	391
N	94	964

Source: National Academies, DataRAP Database, tabulations by staff.

Long Term

This question focuses on how the award has affected the Research Associate's career to date. This question is a bit problematic, given that it is not clear what an excellent or poor affect is, or whether all Research Associates are thinking in the same terms in answering this question. Forty-one percent (of 253) NIST/NRC Research Associates and 21 percent (of 6,936) of research associates from other RAPs provided information on this category. Among those who responded, as seen in Table 3-7, the scores were similar to those for short term value above, with an average of 8.7. NIST/NRC Research Associates again were slightly lower at 8.5, as compared with the average of 8.8 for Research Associates of other RAPs. This difference was statistically significant at the 0.05 level.[46]

[46] Based on a t-test. Note the unequal sample sizes and that the samples are not normally distributed.

TABLE 3-7 Research Associates' Appraisal of the Long-Term Value of the Research Associateship Program, by Program

On A Scale of 1-10 (Poor-Excellent), Please Rate the Following: Long Term Value	NIST/NRC RAP	Other RAPs
1	1	13
3		11
4	2	7
5	5	52
6	2	29
7	8	109
8	25	278
9	33	330
10	28	650
N	104	1479

Source: National Academies, DataRAP Database, tabulations by staff.

Laboratory Support

Research Associates' appraisal of lab support encompasses a number of issues: equipment, funding, orientation, safety and health guidelines, etc. Forty-two percent (of 253) NIST/NRC Research Associates and 22 percent (of 6,936) of research associates from other RAPs provided information on this category. Among those who responded, as seen in Table 3-8, as before, the scores are quite high, on average at 8.5. NIST/NRC Research Associates averaged 8.4; while Research Associates of other RAPs averaged 8.5. These differences were not statistically significant.

TABLE 3-8 Research Associates' Appraisal of Laboratory Support Research Associateship Program, by Program

On A Scale of 1-10 (Poor-Excellent), Please Rate the Following: Lab Support	NIST/NRC RAP	Other RAPs
1	1	12
2		10
3	1	15
4	1	21
5	7	63
6	3	53
7	12	151
8	19	279
9	24	277
10	38	615
N	106	1496

Source: National Academies, DataRAP Database, tabulations by staff.

Advisor Support

This number of responses declined significantly at this point in the questionnaire. Very few respondents answered this question, which focuses on the quality of mentoring received from the research advisor. This question ignores the possibility that the Research Associate received mentoring from someone else either at the host agency or external to it. Sixteen percent (of 253) NIST/NRC Research Associates and three percent (of 6,936) of research associates from other RAPs provided information on this category. Among those who responded, as seen in Table 3-9, and somewhat surprisingly given the comments of the former Research Associates, the average score was quite high at 8.6, and again NIST/NRC Research Associates were slightly lower than Research Associates of other RAPs (8.2 to 8.7). This difference is statistically significant at the 0.05 level, but should be taken very cautiously. However, given the very low response rate, it is likely that people who received a lot of support might be more likely to answer the question.

TABLE 3-9 Research Associates' Appraisal of the Quality of Mentoring by Their Advisor, by Program

On A Scale of 1-10 (Poor-Excellent), Please Rate the Following: Advisor/Mentor Support	NIST/NRC RAP	Other RAPs
1		1
2	1	1
3	1	1
4	1	2
5	1	7
6	1	8
7	5	14
8	10	43
9	6	42
10	14	93
N	40	212

Sources: National Academies, DataRAP Database, tabulations by staff.

LPR (administrative) Support

"LPR" is an acronym for Laboratory (e.g., NIST) NRC Program Representative. This question is designed to tap the administrative support Research Associates get on-site. Even less respondents answered this question. Fourteen percent (of 253) NIST/NRC Research Associates and three percent (of 6,936) of research associates from other RAPs provided information on this category. Among those who responded, as seen in Table 3-10, and although little should be made of this finding, it was the factor with the lowest average score (7.8). The NIST/NRC Research Associates gave it a 7.7, while the Research Associates of other RAPs gave it an average score of 7.9. This difference is not statistically significant.

TABLE 3-10 Research Associates' Appraisal of Support at Their Host Agency, by Program

On A Scale of 1-10 (Poor-Excellent), Please Rate the Following: LPR Support	NIST/NRC RAP	Other RAPs
0		1
1		4
2		4
3	2	6
4		1
5	4	18
6		6
7	8	27
8	7	51
9	8	26
10	6	67
N	35	211

Source: National Academies, DataRAP Database, tabulations by staff.

NRC Support

Finally, Research Associates are asked to appraise administrative support provided by the NRC. It is the case that all Research Associates have interaction with the NRC, as it administers the RAPs. Interestingly, more respondents answered this question than the questions on support, even though this question appears last on the list. Twenty-six percent (of 253) NIST/NRC Research Associates and 21 percent (of 6,936) of research associates from other RAPs provided information on this category. Among those who responded, as seen in Table 3-11, in general, there was positive feedback from Research Associates with an average score of 8.8. However, in this one instance, there was a larger difference between the NIST/NRC Research Associates and the Research Associates of other RAPs (8.0 to 8.8). This difference is statistically significant at the 0.05 level. There is no evidence as to why the NIST/NRC Research Associates would have a less favorable view of the NRC's administration. This might be a fruitful line of future inquiry by the NRC's Fellowships Office.

TABLE 3-11 Research Associates' Appraisal of the Support of the NRC, by Program

On A Scale of 1-10 (Poor-Excellent), Please Rate the Following: NRC Support	NIST/NRC RAP	Other RAPs
1		2
2	1	6
3	1	6
4	1	10
5	7	45
6	2	31
7	8	118
8	19	267
9	11	326
10	17	676
N	67	1487

Source: National Academies, DataRAP Database, tabulations by staff.

RESEARCH ADVISORS' EVALUATION OF RESEARCH ASSOCIATES

The current evaluation of the associate by the research advisor (see Appendix I) is limited as it is currently not filled out by many advisors, does not go into much depth concerning the value of the program, and is currently designed to tackle to separate objectives: whether the evaluation is for the renewal of their term (where the data are collected by the host agency) or for the end of the term (where the data are collected by the NRC).

PRELIMINARY RESULTS

Currently available data do not allow for a program evaluation of immediate outcomes of the Program. Little data are collected on Research Associates' experiences or on research advisors' evaluation of Research Associates. Data are also not collected on the value of the program to NIST or to the broader scientific and engineering community.

Second, with the caveat that this conclusion is based on very limited data that may be biased by nonresponse, NIST/NRC RAs are as productive as Research Associates in other programs. NIST/NRC RAs, statistically were more likely to receive an award or give domestic presentations than Research Associates in other Programs. Conversely, they published fewer journal articles. However, while these differences were statistically significant, they were not substantively large. NIST/NRC RAs patent or give international presentations comparably with Research Associates in other Programs. Finally, and subject to the same caveat, RAs are quite satisfied with the program. On a scale of 1 to 10, with 10 being excellent, NIST/NRC RAs rated short-term and long-term value of the program; lab, advisor, administrative (NIST and NRC) support between 7.7 and 8.5. In half the categories NIST/NRC RAs and RAs in other programs reported statistically similar levels of satisfaction. In the other half, other RAs reported higher levels of satisfaction.

RECOMMENDATIONS

1. **NIST should conduct a more thorough assessment of RAs' experiences during the postdoctoral appointment, their satisfaction with and views on the benefits of the Program, and NIST staff's satisfaction with and views on the benefits of the Program.**
    a. To assist in this, the NRC should redesign the final report and the Research Advisor's evaluation form to maximize the collection of data from these instruments (see Box 3-1 and Box 3-2 for suggested questions).
    b. The final report and the Research Advisor's evaluation should be made mandatory.
    c. Some elements of the current data collected could be subjected to further analysis.
        i. For example, NIST may wish to conduct further analysis on peer-reviewed journals, for example by:
            1. asking whether the RA was sole or lead author,
            2. examining whether RAs publish with NIST staff, and
            3. examining the quality of the journals in which RAs publish, although this requires some ranking of journals.
        ii. NIST may wish to conduct an impact analysis of RAs' productivity, for example by:
            1. conducting a citation analysis to see how often RAs' publications are referenced by others (note this can be accomplished using citation indexes), or
            2. assessing the type or size of grants postdocs receive.
        iii. NIST may wish to conduct a more thorough review of their support of RAs, asking how familiar they are with NIST administrative offices, how often they turn to those offices for help, and for what reasons.
    d. NIST could also conduct a social network analysis of the collaboration of the RAs (or of NIST employees) to see how the RAPs facilitates new or wider collaboration among scientists and engineers.
    e. When data allow, NIST could consider disaggregating productivity and satisfaction measures for RAs by lab, gender, and race/ethnicity.

Box 3-1
Suggested Final Report for Research Associates

1. Name
2. Contact Information
    a. Address
    b. Phone
    c. Email
3. Information about the postdoctoral appointment
    a. Agency name
    b. Laboratory or center name
    c. Division/Directorate/Department
    d. Postdoctoral start date
    e. Postdoctoral end dates
    f. Name of advisor
    g. Title of research proposal
    h. Summary of research
    i. Relationship of research conducted to research proposal
        i. I did what I proposed
        ii. I did what I proposed, and also did other research projects
        iii. I did some of what I proposed and also did other research
        iv. I did not do what I proposed
4. Was the agency where you undertook the RAP your first choice?
    a. If no, why not: _____________
5. What was your primary reason for taking this postdoc?
    a. Additional training in Ph.D. field
    b. Training in an area outside of Ph.D. field
    c. Work with a specific person or place
    d. Other employment not available
    e. Postdoc generally expected for a career in this field
    f. Salary/benefits
    g. Location
    h. Only offer received
    i. Some other reason: ______________
6. When you applied to the RAP, did you apply to multiple agencies? Which other ones did you apply to?
7. Around the time you applied to the RAP, did you apply to other postdoctoral positions?
8. Were you offered more than one position?

Box 3-1 (continued)
Suggested Final Report for Research Associates

9. If yes, why did you choose the NIST postdoc?
    a. Stipend was better
    b. Prestige of agency
    c. Stepping stone to career
    d. Salary/benefits
    e. Location
    f. Other: ____________
10. In addition to conducting research, which of the following professional or career development activities did you engage in during your postdoctoral appointment?
    a. Guest lecturing at host institution
    b. Advising/mentoring others at host institution
    c. Organizing seminars or workshops
    d. Attending workshops, lectures, seminars in your research area
    e. Attending seminars on proposal writing/grant making
    f. None of the above
    g. Other: ______________
11. To what extent did the postdoctoral appointment … (all should be on 1-5 scale)
    a. Increase you subject matter knowledge or expertise (great extent, somewhat, not at all)
    b. Improve specific research skills or techniques
    c. Increase contacts with colleagues in your field
    d. Provide opportunities to use specialized equipment
    e. Improve your problem-solving skills
    f. Enhance your career opportunities
    g. Help in other areas: _______________
12. Was the postdoc experience what you expected in terms of…
    a. Ability to conduct your own research
    b. Access to research equipment, facilities, resources
    c. Ability to work independently
    d. Ability to collaborate/network with others at the agency
    e. Ability to collaborate/network with others outside the agency
    f. Ability to network with other postdocs
    g. Mentoring or advising
    h. Ability to publish
    i. Ability to apply for grants
    j. Ability to travel to conferences, professional meetings, etc.
    k. Administrative support from agency
    l. Administrative support from NRC

Box 3-1 (continued)
Suggested Final Report for Research Associates

13. In your opinion, what would have been the optimal duration of your postdoctoral appointment? ______ months
14. Outcomes. Please list your…
    a. Publications
        i. Books, book chapters
        ii. Publications in peer-reviewed journals
        iii. Other publications
    b. Patents awarded
    c. Presentations
        i. Domestic
        ii. International
    d. Seminars or lectures delivered
    e. Awards received
    f. Grants
15. Thinking about your career plans when you began the postdoc and now, has your postdoctoral experience had an effect on your career preferences? Would you say that today you are more likely, about as likely, or less likely to work in:
    a. Government
    b. Industry
    c. Academia
    d. Nonprofit
    e. Self employed
16. Where are you planning to go next?

Box 3-1 (continued)
Suggested Final Report for Research Associates

17. Post-postdoc career plans
    a. What are your current plans?
        i. Looking for another postdoc
        ii. Looking for employment
        iii. Employed
        iv. Not looking for employment
    b. If employed…
        i. Position title
        ii. Employer name
        iii. Employer type
            1. Remain at host agency as permanent employee
            2. Remain at host agency as contract/temporary employee
            3. Other government position
            4. Academic position
            5. Industry
            6. Nonprofit
            7. Self employed
            8. Other: ____________
18. What were the best features of the postdoc?
19. What were the worst features of the postdoc?
20. Have you recommended the postdoc to others?
21. If you could make improvements to the program what would they be?

Box 3-2
Suggested Research Advisor Evaluation

1. Name
2. Contact Information
   a. Address
   b. Phone
   c. Email
3. Have you ever been an advisor to a postdoc? (If yes, skip to 5)
4. If no, why not? (Continue to 10)
5. If yes, how many in the past five years?
6. Do you keep in touch with former postdocs you advised?
7. Are you currently an advisor to a postdoc?
8. If yes, how would you rate the postdoc associate to other comparable employees at your agency?
   a. Knowledge of field (below ave, ave, above ave, good, exceptional)
   b. Research technique
   c. Motivation/initiative
   d. Independent research
   e. Innovative thinking
   f. Overall scientific ability
9. Would you like the postdoc as a professional colleague at your agency?
10. Should the postdoc program be limited to U.S. citizens (if no, who else should be allowed to apply?)
11. How do you think most doctorates hear about the postdoctoral program at your agency?
    a. Word of mouth from fellow graduate students/doctorates
    b. Ph.D. thesis advisor or other professor
    c. University placement office
    d. Former or current postdoc with your agency
    e. Agency employee
    f. NRC presentation at professional meeting
    g. Advertisement in professional publication
    h. Internet
    i. Other: _______________
12. What type of outreach do you think is most effective?
13. In your opinion, is 2 years the optimal duration for the postdoctoral appointment?

Box 3-2 (continued)
Suggested Research Advisor Evaluation

22. To what extent do you agree or disagree with the following statements? (Have them use 1-5 scale)
    a. Program increases postdocs' knowledge of field
    b. Program allows postdocs to learn new fields
    c. Program allows postdocs to Become more interdisciplinary
    d. Program allows postdocs to Try out working at agency
    e. Program allows postdocs to Improve research techniques
    f. Program allows postdocs to practice prepare grants
    g. Program allows postdocs to Practice giving presentations
    h. Program allows postdocs to Publishing
    i. Program allows postdocs to Collaboration with agency employees
    j. Program allows postdocs to Collaboration with others outside agency
    k. Program makes postdocs more independent as researchers
    l. Program makes postdocs more innovative thinkers
23. To what extent do you agree or disagree with the following statements?
    a. My agency has increased collaboration due to the postdoc program
    b. My agency is able to cover a wider range of research topics because of the postdocs
    c. My agency is able to get more research done because of the postdocs
    d. The postdoc program increases the applicant pool for regular appointments at my agency
24. What benefits did you get from being an advisor?
25. Have you made any efforts to recruit postdocs? If yes, did you
    a. Give presentations at universities
    b. Meet with graduate students/doctorates at universities
    c. Meet with graduate students/doctorates at professional meetings
    d. Invite graduate students/doctorates to visit or give presentations at your agency
    e. Other: ________________
26. Does your group make any efforts to recruit postdocs
    a. If yes, what?
27. Does your lab or center make any efforts to recruit postdocs?
    a. If yes, what?
28. What were the best features of the postdoc?
29. What were the worst features of the postdoc?
30. If you could make improvements to the program what would they be?

4

Careers

There is very little information on the careers of former NIST/NRC RAs. Sources of information are described in the next section, but to summarize, the data cover the immediate post-appointment plans of RAs or their current employment at an arbitrary, recent time point. Career histories are not available. Based on these data, former RAs are seen to have moved into employment in all sectors. Of particular interest to NIST is whether the NRC Resident Research Associateship Program at NIST (NIST/NRC RAP) is providing a source of job candidates. Some information available to answer this question was obtained through the expert panels and data collected by NIST.

From the point of view of current and former RAs at NIST, as well as NIST staff, the Program is viewed as beneficial to expanding the pool of potential applicants to NIST jobs. Many current RAs, interviewed during the expert panels came to the RAP with the hopes of remaining at NIST. They were aware that it was difficult to make the transition to permanent employee, but some felt that having the NIST/NRC Research Associateship was the best way to stay. The success rate, based on their estimates, varies widely by where at NIST the RAs were employed. Estimates ranged from 5 to 60 percent, although the NIST-wide estimate was thought to be around 30 to 40 percent. Former RAs who remained at NIST thought that the overall retention rate was around 50 percent, and that the percentage retained varied by labs, for example in some labs they felt it was 15 to 20 percent. While both current and former RAs noted that it was difficult to remain at NIST, former RAs felt that current RAs overestimate their chances of staying. Advisors and division chiefs put the percentage of RAs being hired at NIST at about 33 percent. They reported that about half of RAs ask about staying. Participants noted that in many cases, RAs intended to go elsewhere after their tenure, while those who came to NIST and highly enjoyed working there often found a way to remain. During the expert panel with NIST advisors and managers, they noted that the program was a good way to recruit (some said "primary" way to recruit) and retain good people. As one participant noted: "It's a great way to try someone out."

RESEARCH ASSOCIATES' CAREERS

Three sources of information about the careers of RAs after the appointment are the final reports RAs fill out at the end of their postdoctoral appointment, a directory put together by The National Academies, and information from NIST about change in status of RAs. In 1996, The National Academies put together a *Directory of Resident Research Associates*. The first data source focuses on the plans of RAs after completing their appointments. It is the only career information taken at the beginning of the RAs post-appointment career.

Table 4-1 looks at where RAs planned to go next, following their appointment. Again, these data are taken from the final reports filled out by RAs, and as noted in Chapter 3, not many RAs complete this form. Fifty-seven percent (of 253) NIST/NRC RAs and 37 percent (of 6,936) of RAs from other RAPs provided information on their post-tenure employment position.

TABLE 4-1 Immediate Employment of Research Associates Following Postdoctoral Appointment, by Program

Plans	NIST/NRC RAP		Other RAP	
	N	%	N	%
Research or teaching at U.S. college/university	28	19.3	609	23.9
Research position at another U.S. govt. agency	10	6.9	247	9.7
Administrative position at U.S. govt. lab	0	0.0	53	2.1
Research/administration in non-profit	1	0.7	43	1.7
Research/administration in industry	24	16.6	325	12.8
Self employed	0	0.0	45	1.8
Postdoctoral research	6	4.1	173	6.8
Remain at host agency as permanent employee	45	31.0	392	15.4
Remain at host agency as contractor/temp	20	13.8	180	7.1
Research or teaching at foreign govt. lab	0	0.0	48	1.9
Research or teaching at foreign college/university	1	0.7	67	2.6
Government	1	0.7	1	0.0
Other	3	2.1	88	3.5
No information provided	3	2.1	188	7.4
Unknown	3	2.1	87	3.4
N	145	100	2546	100

Source: National Academies, DataRAP Database, tabulations by staff.

As Table 4-1 shows, among those RAs who answered the question, the most common response for NIST/NRC RAs was remaining at NIST as a permanent employee. When adding in those RAs who were going to continue working at or for NIST as a contractor or as a temporary employee, about 45 percent of those who answered the question continued to be affiliated with NIST, compared with only 22.5 percent for the Research Associates who were in other RAPs. This lends credence to the views expressed by NIST/NRC Research Associates about their satisfaction with the program, to the view expressed by participants in the expert panels that the Research Associateships are a good entrée into a career at NIST, and to the view that NIST uses the Research Associateships as one recruiting tool for finding skilled employees. Respondents to the final report also are asked to identify the name of the organization that they plan to go to for their next position. For academic appointments, respondents listed 27 different institutions. Aside from NIST, Research Associates identified a number of other government positions, including several at the national labs. As was true for academia, respondents whose immediate employment plans were in industry each cited a different company. Thus, those respondents who were not continuing in some fashion at NIST tended to go all over the country within the major employment sectors.

The *Directory* was intended to identify where former RAs were at the time the *Directory* was compiled. It initially covered the years 1959-1995. The data were then updated to cover up to the period 2002 and made web-accessible. Information in the written Directory and its subsequent update were based on data collected at the time individuals received their reward and responses to a questionnaire requesting information about their current activities. Many former RAs did not respond to the questionnaire and in many cases it was not possible to find contact information for some former RAs (NRC, 1996).

Beginning in 1965, there are 9,924 entries in the current database. Of these, 1,035 or about 10 percent were individuals who had received NIST/NRC Research Associateships. The response rate for current employment data was 37.5 percent (of 1035) for NIST/NRC RAs and 37.7 percent (of 8889) for former RAs in other RAPs. Among those who responded, as Table 4-3 shows, the current employer for over one-third of former NIST/NRC RAs is NIST. NIST/NRC RAs were more likely to be employed in government and less likely to be employed in academia or other sectors. An important finding is that about 37.6 percent of NIST/NRC RAs were working at NIST when surveyed.

TABLE 4-2 Current Employment of Former Research Associates, by Program

	NIST/NRC RAP		Other RAP	
Current Employer of Former Research Associates	N	%	N	%
Academic institution	114	29.4	1186	35.4
Industry	80	20.6	710	21.2
Government	13	3.4	134	4.0
National lab	23	5.9	63	1.9
Government-same as postdoc	146	37.6	941	28.1
Nonprofit	4	1.0	89	2.7
Medical center/Hospital	4	1.0	126	3.8
Other (including self employed)	4	1.0	105	3.1
N	388	100.0	3354	100.0

Source: Fellowship Directory Database.

As Table 4-2 notes—similarly with Table 4-1—among those who answered the questionnaire sent out to compile the Directory, 37.6 percent of former NIST/NRC RAs were employed at NIST when the survey was taken, compared with about 28.1 percent of former RAs of other RAPs, who were at their host agencies.

A final data source are data collected by NIST of changes in postdoctoral status, that is: appointment start dates, appointment terminations, resignations, and most important for our purposes here, conversions of RAs to either term appointments of career conditional appointments.

TABLE 4-3 Number of Former NIST/NRC Research Associates Who Converted to Career-Conditional or Term Appointments After Their Postdoctoral Appointments

Year	Awards	Conversion to Career Conditional	Conversion to Term Appointment	Total
1998	57	1	7	8
1999	40	7	7	14
2000	58	4	7	11
2001	39	4	9	13
2002	57	5	9	14
2003	35	6	6	12
2004	52	3	10	13
2005	56	5	17	22
2006	48	3	11	14
2007	59	1	3	4
Total	501	39	86	125

Note: Awards is for 2 years prior to year of change.
Source: Data provided by NIST, tabulations by staff.

As Table 4-3 shows, about 25 percent of former RAs converted (this does not include RAs who converted to contractor status) and among those former RAs who converted, about 31 percent converted to permanent employee status.

There are a number of reasons to conduct a more thorough evaluation of the careers of former RAs. Collecting data currently not being done on the careers of former RAs would facilitate benchmarking should NIST want to make changes to the program, further improvements to the program, and would shift information about the program from qualitative to quantitative data. There are several directions that future assessment could go in. In general, these include: (1) studying benefits to RAs (potential benefits could include: better or more training, greater productivity after the postdoctoral appointment, receiving more grants after the postdoctoral appointment); (2) studying benefits to NIST (potential benefits could include: success of NIST/NRC RAs, more and better alternatives for hiring, increased breadth of expertise, novel research projects and their impact, increased collaboration, increased numbers of applicants to the program, or increased prestige); and (3) examining the costs of the program to NIST (potential costs could include: costs to advisors or opportunity cost of not hiring other staff). Methodologically, such analysis could take the form of surveys of former RAs, social network analysis (to examine collaboration), CV analysis (to examine the impact of the position on careers), or citation analysis (to assess the impact of RAs' work).

PRELIMINARY RESULTS

Preliminary evidence suggests that RAs contribute to the pool of qualified applicants to permanent positions at NIST. About 45 percent of RAs indicated that their immediate post-tenure position was at NIST as a permanent, temporary, or contract employee after their appointment—a higher percentage than RAs at other federal agencies. A survey of former RAs found that a higher percentage of former NIST/NRC RAs stayed at NIST than RAs at other federal agencies stayed at their host agency (37.6 to 28.1 percent). Second, evidence on the outcomes of the Program is largely lacking. Little data are collected on the career outcomes of former RAs; and the value of the program to NIST or to the broader scientific and engineering community.

RECOMMENDATIONS

1. **NIST should conduct a broad evaluation of the careers of former RAs to evaluate the impact of the Program on RAs' careers, NIST, and the broader science and engineering community.** The best approach for doing this is a survey, which would compare the career outcomes of NIST/NRC RAs to similar postdocs. The survey would be directed towards these former RAs and a suitable control group. Ideally, two possible comparisons could be made. First, one could construct a peer group. This would consist of a matched or stratified sample of individuals who had postdocs similar to the one at NIST for the comparison group. Although not ideal, one solution would be to take a stratified sample of former RAs from the Fellowships Office's Directory. This is a census of former RAs; but as noted earlier in the report, many of these individuals could not be found or failed to respond to an earlier survey designed to collect information on their current employment. A second comparison group would consist of similar doctorates. A roster could be assembled by tapping the group of applicants to RAPs, who did not receive an award. These individuals will likely exhibit a diversity of career paths, including some who took postdocs (in academia or industry) and others who went straight into employment. Box 4-1 offers suggested questions that might be asked.

Box 4-1
Career Assessment Survey of Former Research Associates

1. Which Research Associateship Program were you in?
2. Which lab?
3. Which division/directorate/department?
4. Start date of postdoc
5. End date of postdoc
6. What was your primary reason for taking this postdoc?
    a. Additional training in Ph.D. field
    b. Training in an area outside of Ph.D. field
    c. Work with a specific person or place
    d. Other employment not available
    e. Postdoc generally expected for a career in this field
    f. Salary/benefits
    g. Location
    h. Some other reason: ______________
7. Demographic information
    a. Highest degree
    b. Year received highest degree
    c. Ph.D. field
    d. Gender
    e. Race/ethnicity
    f. Citizenship
8. Employment characteristics
    a. Have you been employed in any position since completing your postdoc?
    b. Are you currently employed (or self-employed) either full-time or part-time?
    c. Where are you currently employed?
        i. Educational institution
        ii. Industry
        iii. Government
        iv. Not-for-profit
        v. Self-employed
        vi. Other: _____________
    d. If educational, are you employed at:
        i. K-12
        ii. Two-year college, community college, or technical institute
        iii. Four-year college or university

Box 4-1 (continued)
Career Assessment Survey of Former Research Associates

9. If government, are you at the same agency that you had the postdoc with?
10. Is your current position a postdoctoral appointment?
11. Is your current employer the same as your first post-postdoctoral appointment employer?
12. If no, what was the type of employer for your first post-postdoctoral appointment employer?
13. Was your first post-postdoctoral appointment employer the type of employer that you envisioned when you applied for the postdoctoral appointment?
14. How useful did you find the following factors in seeking your first post-postdoctoral appointment employer? (1 = not at all useful to 5 = extremely useful, plus Not Applicable)
    a. Contacts initiated by the employer
    b. Contacts you initiated
    c. Contacts provided by your research advisor
    d. Contacts provided by someone else at the agency
    e. Prestige of the your advisor
    f. Prestige of the agency
    g. Prestige of the RAP
    h. Publications for which you received credit while a postdoc
    i. Presentations you gave while a postdoc
    j. Participation of grants
    k. The area you were researching while a postdoc
    l. Other: ___________________
15. Productivity (Over the past five years or since the end of your postdoc)
    a. Publications
        i. Books, book chapters
        ii. Publications in peer-reviewed journals
    b. Patents awarded
    c. Presentations
        i. Domestic
        ii. International
    d. Awards received
    e. Grants

Box 4-1 (continued)
Career Assessment Survey of Former Research Associates

16. Views about how the program helped you (1 = not at all to 5 = a great deal, plus Not Applicable)
    a. To what extent have you continued to stay in touch with various persons you met during the postdoc appointment?
        i. Your research advisor
        ii. Other agency staff
        iii. Other postdocs
17. To what extent do you agree or disagree with the following statements (1 = strongly disagree to 5 = strongly agree)?
    a. Overall, I found my postdoc experience to be valuable
    b. When it came to securing my first post-postdoc position, my postdoc experiences put me on an equal footing with other postdoctoral researchers of similar qualifications
    c. My postdoc experiences taught me most of what I needed to know to prepare grant proposals
    d. My postdoc experiences taught me most of what I needed to know to conduct independent research
    e. My postdoc experience led to a professional expertise that I would not have developed otherwise
    f. I established on-going friendships with people I met at my host institution
    g. I am proud to have been a NIST Postdoc
18. To what extent did your postdoctoral experience hinder or help with each of the following? (1 = no help at all to 5 = extremely helpful, plus Not Applicable)
    a. Quality of your current research
    b. Specific direction of your current research
    c. Progress of your current research
    d. Your success in obtaining subsequent funding
    e. Your teaching/curricular activities
    f. Your ability to mentor others
    g. Your confidence in performing leading-edge research
    h. Your career as a whole
    i. Other: ____________
19. Other
    a. Have you recommended the postdoc to others?
    b. Would you?
    c. What were the best features of the postdoc?
    d. What were the worst features of the postdoc?
    e. If you could make improvements to the program what would they be?

5

Preliminary Results and Recommendations

PRELIMINARY RESULTS

An overall conclusion of the report is that some of the data that would support a full-fledged evaluation of the NIST/NRC RAP are simply not collected at this time. Some data are collected, but the number of RAs and advisers filling out the forms remain small, which may mean that the results based on information provided by those who did fill out the form are not representative of all RAs or all advisers. Thus caution must be exercised in reading the results. With those caveats firmly in mind, there are a number of interesting findings.

Turning first to examine applicants, the application form is a very useful data collection instrument. Among the three current instruments—application form, final report, adviser's evaluation—the application form has produced the most data. Further, personal communication is the most likely means by which applicants hear about the RAPs, including the NIST/NRC RAP. Key findings regarding how applicants heard about the program were:

- Applicants to the NIST/NRC RAP were twice as likely as applicants to the other RAPs to hear about the position initially from their Ph.D. advisor or other professor and somewhat more likely to hear about the program from colleagues or fellow graduate students, but less likely to hear about the program from a research advisor or other scientific staff at the federal laboratory.
- The most common sources of information for applicants to the NIST/NRC RAP were professors or colleagues.
- Male and female applicants heard about the NIST/NRC RAP similarly, except via presentations at professional meetings, which women cited twice as often as men.
- There were no differences by race/ethnicity in how applicants to the NIST/NRC RAP heard about it.

Outreach efforts produce more qualified applicants than NIST has slots to fill for RAs; and the pool of applicants includes many from top research institutions and is increasingly diverse. Overall, 22 percent of applicants were awarded an appointment—a lower ratio than for RAPs elsewhere. Women are increasingly applying to the NIST/NRC RAP and being awarded research associateships. The NIST/NRC RAP seems to be as popular as the other RAPs for women. Underrepresented minorities are increasingly applying to the NIST/NRC RAP and being awarded research associateships. For applicants to the NIST/NRC RAP and awardees, at least half came from 20 of the top doctoral-granting institutions in the United States.

Applicants and awardees to the NIST/NRC RAP differ from their counterparts in the other RAPs. Since 1990, underrepresented minorities are proportionately more likely to be awarded a NIST/NRC Research Associateship than another research associateship. Applicants to, and awardees of the NIST/NRC RAP are younger on average than those who apply for the other programs. NIST/NRC RAP applicants and awardees are more likely to be single. They are more likely to have Ph.D.s in the physical sciences than biological. The majority of awards go to doctorates from the physical sciences. But, because there are so many applications from this discipline, only about one in five applicants with this background receive awards.

Preliminary analysis suggests that labs receive different amounts of applications and awards are not made uniformly across different labs. Some awardees do decline NIST/NRC Research Associateships, though the percentage of declined offers is often lower than that for the other RAPs and has declined over time.

Turning now to an assessment of the experiences of Research Associates, currently available data do not allow for a program evaluation of immediate outcomes of the Program. Little data are collected on Research Associates' experiences or on research advisors' evaluation of RAs. Data are also not collected on the value of the program to NIST or to the broader scientific and engineering community.

Second, with the caveat that this conclusion is based on very limited data that may be biased by nonresponse, NIST/NRC RAs are as productive as RAs in other Programs. NIST/NRC RAs statistically were more likely to receive an award or give domestic presentations than RAs in other Programs. Conversely, they published fewer journal articles. However, while these differences were statistically significant, they were not substantively large. NIST/NRC RAs patent or give international presentations comparably with RAs in other Programs. Finally, and subject to the same caveat, RAs are quite satisfied with the Program. On a scale of 1 to 10, with 10 being excellent, NIST/NRC RAs rated short-term and long-term value of the program; lab, advisor, administrative (NIST and NRC) support between 7.7 and 8.5. In half the categories NIST/NRC RAs and Research Asssociates in other programs reported statistically similar levels of satisfaction. In the other half, other RAs reported higher levels of satisfaction.

Finally, looking at the careers of former RAs, preliminary evidence—which is quite limited—suggests that RAs contribute to the pool of qualified applicants to permanent positions at NIST. About 45 percent of RAs reported that their immediate post-tenure position was at NIST as a permanent, temporary, or contract employee after their appointment—a higher percentage than RAs at other federal agencies. A survey of former RAs found that a higher percentage of former NIST/NRC RAs stayed at NIST than RAs at other federal agencies stayed at their host agency (37.6 to 28.1 percent). Second, evidence on the outcomes of the Program is largely lacking. Little data are collected on the career outcomes of former RAs; and the value of the program to NIST or to the broader scientific and engineering community.

RECOMMENDATIONS

1. **NIST should conduct a more thorough evaluation of the NIST/NRC Research Associateship Program.**
    a. As a first step, NIST and the NRC should review specific goals of the program.
    b. The evaluation should include the following components: an assessment of outreach to potential applicants; an assessment of individuals who decline to accept a Research Associate position; an assessment of the benefits of the program on the RAs after they complete their appointments; an assessment of the benefits to NIST of hosting RAs; and an assessment of benefits of the Program to the broader scientific and engineering community.
2. **NIST should conduct an evaluation of outreach efforts.**
    a. To conduct such an evaluation, data need to be collected. In this regard, the question on the application about how applicants hear about the program is

helpful and should be retained. However, the "Other" category should be further analyzed and a choice of "Website" should be added as a category.

b. Additional data could be collected from NIST personnel and former or current NIST RAs. Such data could be used to answer such questions as:
    i. What mechanisms do NIST personnel and RAs use to interact with potential applicants and
    ii. Which mechanisms seem to work best?
    iii. Has there been any effort to focus specifically on diversity? How?

   Such research could be undertaken via a combination of expert panels or surveys of NIST staff and current or former RAs to answer the first and third questions and to provide information for an assessment of the second question. Information should also be collected on the costs for individual outreach efforts (e.g., money spent on advertisements, time spent meeting with graduates) to compare to the benefits (how many applicants come from each individual outreach type).

c. A second step to facilitate an evaluation of outreach efforts is to identify metrics for quantifying value obtained from different outreach strategies, such as hits to the website or number of graduate students met with at professional meetings.

d. Examine individual outreach strategies for return on investment. This could include such strategies as assessing the NIST website for usability and informational content or assessing the return on advertising in publications. As part of the assessment of the NIST website, NIST could consider adding contact information for Research Advisors to facilitate a dialogue between potential applicants and relevant NIST staff.

e. Finally, consider whether there might be other outreach strategies that are being underused currently, and which might have potential value, such as direct mail to deans, department heads and other university administrators.

f. In addition, it is important to determine if any groups of graduate students—and potential applicants—who would make good candidates for the NIST/NRC RAP are unaware of the Program and how one applies. It would be difficult to craft a random sample of graduate students, but a limited survey might be possible.

3. **NIST should conduct an evaluation of individuals who decline offers of Research Associateships.** This could be done as a telephone interview or via a survey. As there are only a few people who decline each cycle, the burden would be relatively small. Two basic questions should be asked of those who are awarded but decline: (1) why are you declining, and (2) what are you planning to do instead?

4. **The NRC should amend the application form.** The number of fields should be reduced, in particular by collapsing very similar labels and by removing labels that are for multiple fields (e.g., "Biophysics Physics Biochemistry"). At least with regard to Ph.D. fields, an example of a smaller field list is found in the NSF's Survey of Earned Doctorates (see Appendix B).

5. **The NRC should update the DataRAP database to replace organizational names (e.g., institutes or labs) that no longer exist at NIST with current equivalents.**

6. **NIST should conduct a more thorough assessment of RAs' experiences during the postdoctoral appointment, their satisfaction with and views on the benefits of the Program, and NIST staff's satisfaction with and views on the benefits of the Program.**

a. To assist in this, the NRC should redesign the final report and the Research Advisor's evaluation form to maximize the collection of data from these instruments (see Box 3-1 and Box 3-2 for suggested questions).
b. The final report and the Research Advisor's evaluation should be made mandatory.
c. Some elements of the current data collected could be subjected to further analysis.
    i. For example, NIST may wish to conduct further analysis on peer-reviewed journals, for example by:
        1. asking whether the RA was sole or lead author,
        2. examining whether RAs publish with NIST staff, and
        3. examining the quality of the journals in which RAs publish, although this requires some ranking of journals.
    ii. NIST may wish to conduct an impact analysis of RAs' productivity, for example by:
        1. conducting a citation analysis to see how often RAs' publications are referenced by others (note this can be accomplished using citation indexes), or
        2. assessing the type or size of grants postdocs receive.
    iii. NIST may wish to conduct a more thorough review of their support of RAs, asking how familiar they are with NIST administrative offices, how often they turn to those offices for help, and for what reasons.
d. NIST could also conduct a social network analysis of the collaboration of the RAs (or of NIST employees) to see how the Research Associateship Program facilitates new or wider collaboration among scientists and engineers.
e. When data allow, NIST could consider disaggregating productivity and satisfaction measures for RAs by lab, gender, and race/ethnicity.

7. **NIST should conduct a broad evaluation of the careers of former RAs to evaluate the impact of the Program on RAs' careers, NIST, and the broader science and engineering community**. The best approach for doing this is a survey, which would compare the career outcomes of NIST/NRC RAs to similar postdocs. The survey would be directed towards these former RAs and a suitable control group. Ideally, two possible comparisons could be made. First, one could construct a peer group. This would consist of a matched or stratified sample of individuals who had postdocs similar to the one at NIST for the comparison group. Although not ideal, one solution would be to take a stratified sample of former RAs from the Fellowships Office's Directory. This is a census of former RAs; but as noted earlier in the report, many of these individuals could not be found or failed to respond to an earlier survey designed to collect information on their current employment. A second comparison group would consist of similar doctorates. A roster could be assembled by tapping the group of applicants to RAPs, who did not receive an award. These individuals will likely exhibit a diversity of career paths, including some who took postdocs (in academia or industry) and others who went straight into employment.[47]

[47] An alternative approach is to construct a comparison group for the NSF's Survey of Doctorate Recipients by identifying a group of former postdocs. For an example of a report that uses this approach, see: Oak Ridge Institute for Science and Education, 2003.

Bibliography

Austin, J. 2006. Salary survey: U.S. life scientists report rising salaries and high job satisfaction. *Science* 314(5800):842-847.

Bhatiacharjee, Y. 2006. NSF wants PIs to mentor their postdocs. *Science* 313(5788):748.

Curry, E. 2004. *50th Anniversary of the NIST Postdoctoral Research Associates Program.* [Online]. Office of International and Academic Affairs, NIST. Available at: http://www.nist.gov/oiaa/pdanniv.htm.

Davis, G. 2005. Doctors without orders. [Online]. *American Scientist* 93(3, supplement). Available at: http://postdoc.sigmaxi.org/results/.

Ferber, D. 1999. Getting to the front of the bus. *Science* 285(5433):1514-1517.

Finkelstein, N. and J. Libarkin. (no date). *Who Cares About Postdocs Anyway? Evaluating the National Science Foundation's Postdoctoral Fellowships in Science, Mathematics, Engineering and Technology Education.* [Online]. Available at: http://lchc.ucsd.edu/nfinkels/higher_ed_postdoc.pdf.

Hill, S.T., T.B. Hoffer, and M.J. Golladay. 2004. *Plans for Postdoctoral Research Appointments Among Recent U.S. Doctorate Recipients.* NSF InfoBrief, NSF 04-308. Arlington, VA.

Mantovani, R., M.V. Look, and E. Wuerker. 2006. *The Career Achievements of National Research Service Award Postdoctoral Trainees and Fellows: 1975–2004.* Bethesda, MD: ORC Macro.

McCullough, J. and D. Thurgood. 2004. *Outcomes and Impacts of the National Science Foundation's Program of Minority Postdoctoral Research Fellowships.* Arlington, VA: SRI International.

Mervis, J. 2007. Program evaluation: Researchers fault U.S. report critiquing education programs. *Science* 316(5829):1267.

Mervis, J. 1999. Minority postdocs are rare, independent breed. *Science* 285(5433):1529-1530.

Michie, J., X. Zhang, J. Wells, L. Ristow, G. Pion, A. Miyaoka, J. Frechtling. 2007. *Feasibility, Design and Planning Study for Evaluating the NIH Career Development Awards: Final Report.* Rockville, MD: Westat.

National Academy of Sciences, National Academy of Engineering, Institute of Medicine.2007. *Rising Above the Gathering Storm: Energizing and Employing America for a Brighter Economic Future.* Washington, DC: National Academies Press.

National Academy of Sciences, National Academy of Engineering, Institute of Medicine. 2000. *Enhancing the Postdoctoral Experience for Scientists and Engineers: A Guide for Postdoctoral Scholars, Advisers, Institutions, Funding Organizations, and Disciplinary Societies.* Washington, DC: National Academy Press.

National Research Council. 2006a. *Enhancing Philanthropy's Support of Biomedical Scientists: Proceedings of a Workshop on Evaluation.* Washington, DC: National Academies Press.

National Research Council. 2006b. *Evaluation of the Markey Scholars Program.* Washington, DC: National Academies Press.

National Research Council. 2005. *Assessment of NIH Minority Research and Training Programs: Phase 3.* Washington, DC: National Academies Press.

National Research Council. 1996. *Directory of Resident Research Associates 1959-1995.* Washington, DC: National Academy Press.

National Science Board. 1998. *The Federal Role in Science and Engineering Graduate and Postdoctoral Education.* Contribution to the Government/University Partnership. NSF97-235. Arlington, VA. http://www.nsf.gov/nsb/documents/1997/nsb97235/nsb97235.htm.

National Science Foundation (NSF), Division of Science Resources Statistics. 2007. *Graduate Students and Postdoctorates in Science and Engineering: Fall 2005.* NSF 07-321, Project Officer, Julia D. Oliver. Arlington, VA.

NSF, Division of Science Resources Statistics. 2006a. *Graduate Students and Postdoctorates in Science and Engineering: Fall 2004.* NSF 06-325, Project Officer, Julia Oliver. Arlington, VA.

NSF, Division of Science Resources Statistics. 2006b. *Graduate Students and Postdoctorates in Science and Engineering: Fall 2003.* NSF 06-307, Project Officer, Julia Oliver. Arlington, VA.

NSF, Division of Science Resources Statistics. 2006c. *U.S. Doctorates in the 20th Century.* NSF 06-319, Lori Thurgood, Mary J. Golladay, and Susan T. Hill. Arlington, VA.

NSF, Division of Science Resources Statistics. 2006d. *Science and Engineering Doctorate Awards: 2005.* NSF 07-305, Project Officer, Susan T. Hill. Arlington, VA.

NSF, Division of Science Resources Statistics. 2005. *Graduate Students and Postdoctorates in Science and Engineering: Fall 2002.*, NSF 05-310, Project Officers, Julia D. Oliver and Emilda B. Rivers. Arlington, VA.

NSF, Division of Science Resources Statistics. 2003. *Graduate Students and Postdoctorates in Science and Engineering: Fall 2001.* NSF 03-320, Project Officer, Joan Burrelli. Arlington, VA.

NSF, Division of Science Resources Statistics. 2002. *Graduate Students and Postdoctorates in Science and Engineering: Fall 2000.* NSF 02-314, Project Officer, Joan Burrelli. Arlington, VA.

NSF, Division of Science Resources Statistics. 2001. *Graduate Students and Postdoctorates in Science and Engineering: Fall 1999*. NSF 01-315, Project Officer, Joan Burrelli. Arlington, VA.

NSF, Division of Science Resources Statistics. 2000. *Graduate Students and Postdoctorates in Science and Engineering: Fall 1998*. NSF 00-322, Project Officer, Joan Burrelli. Arlington, VA.

NSF, Division of Science Resources Statistics. 1999. *Graduate Students and Postdoctorates in Science and Engineering: Fall 1997*. NSF 99-325, Project Officer, Joan Burrelli. Arlington, VA.

NSF, Division of Science Resources Statistics. 1998. *Graduate Students and Postdoctorates in Science and Engineering: Fall 1996*. NSF 98-307, Project Officer, Joan Burrelli. Arlington, VA.

Nerad, M. and J. Cerny. 1999. Postdoctoral patterns, career advancement, and problems. *Science* 285(5433):1533-1535.

Oak Ridge Institute for Science and Education. 2003. *Alexander Hollaender Distinguished Postdoctoral Fellowship Program: Participant Follow-Up*. Oak Ridge, TN: Oak Ridge Institute for Science and Education.

Rutter, J.L., R.S. Rao, I. Bouchelet, G. Dores, H. Green, J. Gulley, J. Koblinski, M. Lundberg, R. Stolzenberg-Solomon, R. Wheelock, W. Williams, T. Wolfsberg, B. Graubard, P. Hartge, S. Wacholder, and J.P. Schwartz. 2003. *Survey of Mentoring Experiences of NIH Postdoctoral Fellows*. Bethesda, MD: National Institutes of Health.

U.S. Department of Education. 2007. *Report of the Academic Competitiveness Council.* Washington, D.C.

U.S. Government Accountability Office. 2005. *Federal Science, Technology, Engineering, and Mathematics Programs and Related Trends*. GAO-06-114. Washington, D.C.

Vogel, G. 1999. A day in the life of a topflight lab. *Science* 285(5433):1531-1532.

Appendix A

COMMITTEE MEMBERS BIOGRAPHICAL INFORMATION

Dr. Isaac C. Sanchez (NAE), Chair, is William J. Murray Endowed Chair in Engineering, Department of Chemical Engineering at The University of Texas. Dr. Isaac C. Sanchez earned his Ph.D. in chemical engineering from the University of Delaware in 1969. He joined the faculty of The University of Texas at Austin in 1988. In 1997, he was elected to the National Academy of Engineering, the nation's highest engineering honor. Sanchez researches properties of polymer liquids, solutions and blends. He attempts to solve problems in polymer science and engineering by studying polymer interfacial phenomena, and how changes in temperature, pressure and volume affect polymers. Sanchez develops models and uses computer simulations to understand polymer solubility and conformation and to understand the role of water in polymer processes.

Dr. Burt S. Barnow is associate director for research and principal research scientist at the Institute for Policy Studies of the Johns Hopkins University. Dr. Barnow received a B.S. in economics from the Massachusetts Institute of Technology and a Ph.D. in economics from the University of Wisconsin at Madison. His work focuses on the operation of labor markets and evaluating social programs, and his current research includes an evaluation of the welfare-to-work program, an evaluation of training programs to train U.S. workers for jobs currently filled with foreign workers who come to the United States on an H-1B visa, and an evaluation of New Hampshire's welfare reforms. Dr. Barnow also teaches program evaluation in the Institute's graduate public policy program and labor economics in the Department of Economics. Before coming to Johns Hopkins, he was vice president of a consulting firm in the Washington, D.C., area. Dr. Barnow served nine years in the Department of Labor, most recently as director of the Office of Research and Evaluation for the Employment and Training Administration. Dr. Barnow is a member of the Board on Higher Education, the Committee for Review of the Title VI and Fulbright-Hays International Education Programs, the Committee on Meeting the Workforce Needs for the National Vision for Space Exploration, and was a member and Vice-Chair of the Committee on Workforce Needs in Information Technology.

Kathryn Newcomer is the Director of the Ph.D. in Public Policy and Administration program and Associate Director of the School of Public Policy and Public Administration at the George Washington University where she teaches public and nonprofit, program evaluation, research design, and applied statistics. She conducts research and training for federal and local government agencies on performance measurement and program evaluation. Dr. Newcomer has published five books, *Improving Government Performance* (1989), *The Handbook of Practical Program Evaluation* (1994, 2004), and *Using Performance Measurement to Improve Public and Nonprofit Programs* (1997), *Meeting the Challenges of Performance-Oriented Government* (2002) and *Getting Results: A Guide for Federal Leaders and Managers* (2005), and numerous articles in journals including the *Public Administration Review*. She was identified as one of the top 25 evaluation experts in the country in 2001 by the American Journal of Evaluation. She is a Fellow of the National Academy of Public Administration, and currently serves on the Comptroller General's Educators' Advisory Panel. She is serving as President of the National Association of Schools of Public Affairs and Administration (NASPAA) for 2006-2007. She has received two Fulbright awards, one for Taiwan (1993) and one for Egypt (2001-2004). Dr. Newcomer earned a B.S. in education and an M.A. in Political Science from the University of Kansas, and her Ph.D. in political science from the University of Iowa.

Dr. Georgine M. Pion is Research Associate Professor, Department of Psychology and Human Development, Peabody College of Vanderbilt University. Dr. Pion's research has focused on career development and human resource policy, particularly as it pertains to the education, training, and employment of scientists and clinical personnel. In addition to training programs, her work has also involved the conduct of large-scale surveys aimed at evaluating peer review in the neurosciences, identifying the factors that affect satisfaction of NIH applicants for research grants, assessing the supply of and demand for faculty in special education, and trends in the education and employment of psychologists and is an Associate Member of the National Academy of Sciences.

Appendix B
Survey of Earned Doctorates

Survey of Earned Doctorates

July 1, 2006 to June 30, 2007

Conducted by

for

Please complete:

First Name	Middle Name	Last Name	Suffix (e.g., Jr.)
Cross Reference: Birth name or former name legally changed			
Name of Doctoral Institution		City or Branch	
Type of Research Doctoral Degree (e.g., Ph.D, Ed.D, etc.)			Date Degree Granted (mm/yyyy)

This information is solicited under the authority of the National Science Foundation Act of 1950, as amended. ALL INFORMATION YOU PROVIDE WILL BE TREATED AS CONFIDENTIAL and used only for research or statistical purposes by your doctoral institution, the survey sponsors, their contractors, and collaborating researchers for the purpose of analyzing data, preparing scientific reports and articles, and selecting samples for a limited number of carefully defined follow-up studies. The last four digits of your Social Security Number are also solicited under the NSF Act of 1950, as amended; provision of it is voluntary. It will be kept confidential. It is used for quality control, to assure that we identify the correct persons, especially when data are used for statistical purposes in Federal program evaluation. Any information publicly released (such as statistical summaries) will be in a form that does not personally identify you. Your response is voluntary and failure to provide some or all of the requested information will not in any way adversely affect you.

The time needed to complete this form varies according to individual circumstances, but the average time is estimated to be 19 minutes. If you have comments regarding this time estimate, you may write to the National Science Foundation, 4201 Wilson Blvd., Arlington, VA 22230, Attention: NSF Reports Clearance Officer. A Federal agency may not conduct or sponsor a collection of information unless it displays a currently valid OMB control number.

INSTRUCTIONS

Thank you for taking the time to complete this questionnaire. Directions are provided for each question.

- If you have not already done so, please PRINT your name on the front cover.
- Please print all responses; you may use either a pen or a pencil.
- When answering questions that require marking a box, please use an "X."

Part A - EDUCATION

A1. What is the title of your dissertation?

☐ *Please mark (X) this box if the title below refers to a performance, project report, or musical or literary composition required instead of a dissertation.*

Title

A2. Please write the name of the primary field of your dissertation research.

Name of Field

Using the list on page 7, choose the code that best describes the primary field of your dissertation research.

Number of Field ☐☐☐

If your dissertation research was interdisciplinary, list the name and number of your secondary field.

Name of Field

Number of Field ☐☐☐

If there were more than two fields, please continue on the back cover of the questionnaire (p. 10).

A3. Please name the department (or interdisciplinary committee, center, institute, etc.) of the university that supervised your doctoral studies.

Department/Committee/Center/Institute/Program

A4. If you received full or partial tuition remission (waiver) for your doctoral studies, was it:

0 ☐ I did not receive any tuition remission
1 ☐ for less than 1/3 of tuition
2 ☐ between 1/3 and 2/3 of tuition
3 ☐ more than 2/3 of tuition, but less than full
4 ☐ full tuition remission

A5. Which of the following were sources of financial support during graduate school?

Mark ALL that apply

a ☐ Fellowship, scholarship
b ☐ Grant
c ☐ Teaching assistantship
d ☐ Research assistantship
e ☐ Other assistantship
f ☐ Traineeship
g ☐ Internship, clinical residency
h ☐ Loans *(from any source)*
i ☐ Personal savings
j ☐ Personal earnings during graduate school *(other than sources listed above)*
k ☐ Spouse's, partner's, or family's earnings or savings
l ☐ Employer reimbursement/assistance
m ☐ Foreign *(non-U.S.)* support
n ☐ Other - Specify

A6. Which TWO sources listed in A5 provided the most support?

Enter letters of primary and secondary sources

1 ☐ Primary source of support
2 ☐ Secondary source of support

☐ *Mark (X) if no secondary source*

A7. When you receive your doctoral degree, how much money will you owe that is directly related to your undergraduate and graduate education?

Mark (X) one in each column

	UNDERGRADUATE		GRADUATE
0	☐ None	0	☐ None
1	☐ $10,000 or less	1	☐ $10,000 or less
2	☐ $10,001 - $20,000	2	☐ $10,001 - $20,000
3	☐ $20,001 - $30,000	3	☐ $20,001 - $30,000
4	☐ $30,001 - $40,000	4	☐ $30,001 - $40,000
5	☐ $40,001 - $50,000	5	☐ $40,001 - $50,000
6	☐ $50,001 - $60,000	6	☐ $50,001 - $60,000
7	☐ $60,001 - $70,000	7	☐ $60,001 - $70,000
8	☐ $70,001 or more - Specify	8	☐ $70,001 or more - Specify
	$		$

2

A8. The next few questions ask about the degrees you have received. Starting with this doctoral degree, please provide the following information for the most recent master's degree and your first bachelor's degree.

	This research doctoral degree	Most recent master's degree (e.g. MS, MA, MBA) or equivalent	First bachelor's degree (e.g. BA, BS, AB) or equivalent
a. Have you received a degree of this type?	X Yes No	Yes No	Yes No
b. Month / year that you started your degree	Month Year	Month Year	Month Year
c. Month / year degree granted	Month Year	Month Year	Month Year
d. Primary field of study			
e. Field number from list on p. 7			
f. Institution name			
g. Branch or city			
h. State or province			
i. Country	USA		

A9. Excluding those above, have you attained any additional postsecondary degrees? Yes No

If yes, please list the additional degree(s), granting institution(s), and years.

Degree Type
Degree Field
Field Number, p. 7
Month/Year Granted
Institution
Branch or City
State or Country

Degree Type
Degree Field
Field Number, p. 7
Month/Year Granted
Institution
Branch or City
State or Country

If necessary, please continue this list on the back cover (p. 10).

A10. Was a master's degree a prerequisite for admission to your doctoral program? Yes No

A11. In what month and year did you first enter graduate school in any program or capacity, in any university? Month Year

A12. How many years were you:

a. taking courses or preparing for exams for this doctoral degree (including a master's degree, if that was part of your doctoral program)? Years *(round to whole years)*

b. working on your dissertation after coursework and exams (non-course related preparation, writing and defense)? Years *(round to whole years)*

A13. Was there any time from the year you entered your doctoral program and the award of your doctorate that you were not working on your degree (that is, not taking courses or working on your dissertation)? Yes No

If yes, please provide the number of years Years *(round to whole years)*

3

A14. Did you earn college credit from a community or two-year college?

1 ☐ Yes 2 ☐ No

A15. Are you earning, or have you earned, an MD or a DDS?

1 ☐ Yes 2 ☐ No

Part B - POSTGRADUATION PLANS

B1. In what country or state do you intend to live after graduation (within the next year)?

0 ☐ in U.S. ⟶ State ______

1 ☐ not in U.S. ⟶ Country ______

B2. Do you intend to take a "postdoc" position?

(A "postdoc" is a temporary position primarily for gaining additional education and training in research, usually awarded in academe, industry, or government.)

1 ☐ Yes 2 ☐ No

B3. What is the status of your postgraduate plans (in the next year)?

Mark (X) one

0 ☐ Returning to, or continuing in, predoctoral employment
1 ☐ Have signed contract or made definite commitment for a "postdoc" or other work
→ GO TO B4

2 ☐ Negotiating with one or more specific organizations
3 ☐ Seeking position but have no specific prospects
4 ☐ Other full-time degree program *(e.g., MD, DDS, JD, MBA, etc.)*
5 ☐ Do not plan to work or study *(e.g., family commitments, etc.)*
6 ☐ Other - Specify ______
→ SKIP TO C1

B4. What best describes your (within the next year) postgraduate plans?

Mark (X) one

"POSTDOC" OR FURTHER TRAINING

0 ☐ "Postdoc" fellowship
1 ☐ "Postdoc" research associateship
2 ☐ Traineeship
3 ☐ Intern, clinical residency
4 ☐ Other - Specify ______
→ GO TO B5

EMPLOYMENT

5 ☐ Employment *(other than "postdoc" or further training)*
6 ☐ Military service
7 ☐ Other - Specify ______
→ SKIP TO B6

B5. What will be the main source of financial support for your "postdoc" or further training within the next year?

Mark (X) one

0 ☐ U.S. government
1 ☐ Industry/business
2 ☐ College or university
3 ☐ Private foundation
4 ☐ Nonprofit, other than private foundation or college
5 ☐ Foreign government
6 ☐ Other - Specify ______
7 ☐ Unknown

B6. What type of principal employer will you be working for (or training with) in the next year?

Mark (X) one

EDUCATION

1 ☐ U.S. 4-year college or university other than medical school
2 ☐ U.S. medical school *(including university-affiliated hospital or medical center)*
3 ☐ U.S. university-affiliated research institute
4 ☐ U.S. community or two-year college
5 ☐ U.S. preschool, elementary, middle, secondary school or school system
6 ☐ Foreign educational institution

GOVERNMENT *(other than education institution)*

7 ☐ Foreign government
8 ☐ U.S. federal government
9 ☐ U.S. state government
10 ☐ U.S. local government

PRIVATE SECTOR *(other than education institution)*

11 ☐ Not for profit organization
12 ☐ Industry or business *(for profit)*

OTHER

13 ☐ Self-employed
14 ☐ Other - Specify ______

B7. Please name the organization and geographic location where you will work or study.

Name .... ______

State *(if U.S.)* ...... ______

Country *(if not U.S.)* . ______

4

B8. What will be your primary and secondary work activities?

Mark (X) one in each column

	PRIMARY	SECONDARY
Research and development	1	1
Teaching	2	2
Management or administration	3	3
Professional services to individuals	4	4
Other - Specify	5	5

Mark (X) if no secondary work activities

Part C - BACKGROUND INFORMATION

C1. Are you -

1 Male   2 Female

C2. What is your marital status?

Mark (X) one

1 Married
2 Living in a marriage-like relationship
3 Widowed
4 Separated
5 Divorced
6 Never married

C3. Not including yourself or your spouse/partner, how many dependents (children or adults) do you have - that is, how many others receive at least one half of their financial support from you?

Mark (X) if none

Write in number

5 years of age or younger
6 to 18 years
19 years or older

C4. What is the highest educational attainment of your mother and father?

Mark (X) one for each parent

	a MOTHER	b FATHER
Less than high/secondary school graduate	1	1
High/secondary school graduate	2	2
Some college	3	3
Bachelor's degree	4	4
Master's degree *(e.g., MA, MS, MBA, MSW, etc.)*	5	5
Professional degree *(e.g., MD, DDS, JD, D.Min, Psy.D., etc.)*	6	6
Research doctoral degree	7	7
Not applicable	8	8

C5. What is your place of birth?

State *(if U.S.)*
OR
Country *(if not U.S.)*

C6. What is your date of birth?

Month __ __ Day __ __ Year 1 9 __ __

C7. What is your citizenship status?

Mark (X) one

U.S. CITIZEN

0 Since birth
1 Naturalized
SKIP TO C9

NON-U.S. CITIZEN

2 With a Permanent U.S. Resident Visa *("Green Card")*
3 With a Temporary U.S. Visa
GO TO C8

C8. *(If a non-U.S. citizen)*
Of which country are you a citizen?

Specify country of present citizenship

C9. In what state or country was the high school/secondary school that you last attended?

State *(if U.S.)*
OR
Country *(if not U.S.)*

C10. Are you a person with a disability?

1 ☐ Yes → GO TO C11

2 ☐ No → SKIP TO C12

C11. Which of the following categories describes your disability(ies)?

Mark (X) one or more

a ☐ Blind/Visually Impaired

b ☐ Deaf/Hard of Hearing

c ☐ Physical/Orthopedic Disability

d ☐ Learning/Cognitive Disability

e ☐ Vocal/Speech Disability

f ☐ Other - Specify

C12. Are you Hispanic or Latino?

1 ☐ Yes → GO TO C13

2 ☐ No → SKIP TO C14

C13. Which of the following best describes your Hispanic origin or descent?

Mark (X) one

1 ☐ Mexican or Chicano

2 ☐ Puerto Rican

3 ☐ Cuban

4 ☐ Other Hispanic - Specify

C14. What is your racial background?

Mark (X) one or more

a ☐ American Indian or Alaska Native

Specify tribal affiliation(s)

b ☐ Native Hawaiian or other Pacific Islander

c ☐ Asian

d ☐ Black or African-American

e ☐ White

C15. Please fill in the last four digits of your Social Security Number.

X X X - X X - ☐ ☐ ☐ ☐

C16. In case we need to clarify some of the information you have provided, please list an e-mail address and telephone number where you can be reached.

E-mail Address

Daytime or Cell Telephone

C17. Please provide your address and the name and address of a person who is likely to know where you can be reached.

YOUR CURRENT ADDRESS:

Street Address

City/State/Country/Zip or Postal Code

CURRENT ADDRESS OF A PERSON WHO WILL KNOW WHERE YOU CAN BE REACHED:

Name

Street Address

City/State/Country/Zip or Postal Code

The results of this survey will be published in a Summary Report; the Summary Reports on earlier surveys are available at http://www.norc.uchicago.edu/issues/docdata.htm.

Please use the back cover to make any additional comments you may have about this survey.

Thank you for completing the questionnaire. Please return this questionnaire to your GRADUATE SCHOOL for forwarding to Survey of Earned Doctorates, NORC at the University of Chicago, 1 N. State Street, Floor 16, Chicago, IL 60602.

If you have questions or concerns about the survey, you may contact us by e-mail at 4800-sed@norc.uchicago.edu or phone at 1-800-248-8649.

BUSINESS MANAGEMENT/ADMINISTRATION

900 Accounting
905 Banking/Financial Support Services
910 Business Administration & Management
915 Business/Managerial Economics
901 Finance
921 Human Resources Development
916 International Business/Trade/Commerce
920 Marketing Management & Research
917 Management Information Systems/Business Statistics
930 Operations Research *(also in ENGINEERING & in MATHEMATICS)*
935 Organizational Behavior *(see also PSYCHOLOGY/Industrial & Organizational)*
938 Business Management/Administration, General
939 Business Management/Administration, Other

COMMUNICATION

940 Communication Research
957 Communication Theory
950 Film, Radio, TV & Digital Communication
947 Mass Communication/Media Studies
958 Communication, General
959 Communication, Other

COMPUTER & INFORMATION SCIENCES

400 Computer Science
410 Information Science & Systems
419 Computer & Information Science, Other

EDUCATION

RESEARCH & ADMINISTRATION

840 Counseling Education/Counseling & Guidance
800 Curriculum & Instruction
805 Educational Administration & Supervision
820 Educational Assessment/Testing/Measurement
810 Educational/Instructional Media Design
807 Educational Leadership
822 Educational Psychology *(also in PSYCHOLOGY)*
815 Educational Statistics/Research Methods
845 Higher Education/Evaluation & Research
825 School Psychology *(also in PSYCHOLOGY)*
830 Social/Philosophical Foundations of Education
835 Special Education

TEACHER EDUCATION

858 Adult & Continuing Teacher Education
852 Elementary Teacher Education
850 Pre-elementary/Early Childhood Teacher Education
856 Secondary Teacher Education

TEACHING FIELDS

860 Agricultural Education
861 Art Education
862 Business Education
864 English Education
870 Family & Consumer/Human Science *(also in Fields Not Elsewhere Classified)*
866 Foreign Languages Education
868 Health Education
874 Mathematics Education
876 Music Education
878 Nursing Education
880 Physical Education & Coaching
882 Reading Education
884 Science Education
885 Social Science Education
888 Trade & Industrial Education
889 Teacher Education & Professional Development

OTHER EDUCATION

898 Education, General
899 Education, Other

ENGINEERING

300 Aerospace, Aeronautical & Astronautical Engineering
303 Agricultural Engineering
306 Bioengineering & Biomedical Engineering
309 Ceramic Sciences Engineering
312 Chemical Engineering
315 Civil Engineering
318 Communications Engineering
321 Computer Engineering
324 Electrical, Electronics & Communications Engineering
376 Engineering Management & Administration
327 Engineering Mechanics
330 Engineering Physics
333 Engineering Science
336 Environmental Health Engineering
339 Industrial & Manufacturing Engineering
342 Materials Science Engineering
345 Mechanical Engineering
348 Metallurgical Engineering
351 Mining & Mineral Engineering
357 Nuclear Engineering
360 Ocean Engineering
363 Operations Research *(also in MATHEMATICS & in BUSINESS MANAGEMENT)*
366 Petroleum Engineering
369 Polymer & Plastics Engineering
372 Systems Engineering
398 Engineering, General
399 Engineering, Other

HUMANITIES

HISTORY

706 African History
700 American History (U.S. & Canada)
703 Asian History
705 European History
710 History, Science & Technology & Society
707 Latin American History
708 Middle/Near East Studies
718 History, General
719 History, Other

FOREIGN LANGUAGES & LITERATURE

768 Arabic
758 Chinese
740 French
743 German
746 Italian
762 Japanese
752 Russian
755 Slavic (other than Russian)
749 Spanish
769 Other Languages & Literature

LETTERS

732 American Literature (U.S. & Canada)
720 Classics
723 Comparative Literature
735 Creative Writing
734 English Language
733 English Literature (British & Commonwealth)
724 Folklore
736 Speech & Rhetorical Studies
738 Letters, General
739 Letters, Other

OTHER HUMANITIES

770 American/U.S. Studies
773 Archaeology
776 Art History/Criticism/Conservation
792 Bible/Biblical Studies
795 Drama/Theater Arts
780 Music
786 Music Theory & Composition
787 Music Performance
788 Musicology/Ethnomusicology
789 Music, Other
785 Philosophy
790 Religion/Religious Studies
798 Humanities, General
799 Humanities, Other

LIFE SCIENCES

AGRICULTURAL SCIENCES/NATURAL RESOURCES

005 Agricultural Animal Breeding
000 Agricultural Economics
025 Agricultural & Horticultural Plant Breeding
020 Agronomy & Crop Science
010 Animal Nutrition
014 Animal Science, Poultry (or Avian)
019 Animal Science, Other
081 Environmental Science
055 Fishing & Fisheries Sciences/Management
043 Food Science
044 Food Science & Technology, Other
066 Forest Sciences & Biology
070 Forest/Resources Management
079 Forestry & Related Science, Other
050 Horticulture Science
074 Natural Resources/Conservation
030 Plant Pathology/Phytopathology *(also in BIOLOGICAL SCIENCES)*
039 Plant Sciences, Other
046 Soil Chemistry/Microbiology
049 Soil Sciences, Other
080 Wildlife/Range Management
072 Wood Science & Pulp/Paper Technology
098 Agriculture, General
099 Agricultural Science, Other

LIFE SCIENCES CONT.

BIOLOGICAL/BIOMEDICAL SCIENCES

130 Anatomy
110 Bacteriology
100 Biochemistry *(see also PHYSICAL SCIENCES/Chemistry, other)*
102 Bioinformatics
103 Biomedical Sciences
133 Biometrics & Biostatistics
105 Biophysics *(also in PHYSICS)*
107 Biotechnology
129 Botany/Plant Biology
158 Cancer Biology
136 Cell/Cellular Biology & Histology
142 Developmental Biology/Embryology
139 Ecology
145 Endocrinology
148 Entomology
137 Evolutionary Biology
170 Genetics/Genomics, Human & Animal
151 Immunology
157 Microbiology
154 Molecular Biology
160 Neurosciences
163 Nutrition Sciences
166 Parasitology
175 Pathology, Human & Animal
180 Pharmacology, Human & Animal
185 Physiology, Human & Animal
115 Plant Genetics
120 Plant Pathology/Phytopathology *(also in AGRICULTURAL SCIENCES)*
125 Plant Physiology
169 Toxicology
189 Zoology
198 Biology/Biomedical Sciences, General
199 Biology/Biomedical Sciences, Other

HEALTH SCIENCES

210 Environmental Health
211 Environmental Toxicology
220 Epidemiology
212 Health Systems/Service Administration
222 Kinesiology/Exercise Science
240 Medicinal/Pharmaceutical Sciences
230 Nursing Science
215 Public Health
245 Rehabilitation/Therapeutic Services
200 Speech-Language Pathology & Audiology
250 Veterinary Sciences
298 Health Sciences, General
299 Health Sciences, Other

MATHEMATICS

425 Algebra
430 Analysis & Functional Analysis
420 Applied Mathematics
460 Computing Theory & Practice
435 Geometry/Geometric Analysis
440 Logic
445 Number Theory
465 Operations Research *(also in ENGINEERING & in BUSINESS MANAGEMENT/ADMIN.)*
450 Statistics *(also in SOCIAL SCIENCES)*
455 Topology/Foundations
498 Mathematics/Statistics, General
499 Mathematics/Statistics, Other

PHYSICAL SCIENCES

ASTRONOMY

500 Astronomy
505 Astrophysics

ATMOSPHERIC SCIENCE & METEOROLOGY

510 Atmospheric Chemistry & Climatology
512 Atmospheric Physics & Dynamics
514 Meteorology
518 Atmospheric Science/Meteorology, General
519 Atmospheric Science/Meteorology, Other

CHEMISTRY

520 Analytical Chemistry
522 Inorganic Chemistry
526 Organic Chemistry
530 Physical Chemistry
532 Polymer Chemistry
534 Theoretical Chemistry
538 Chemistry, General
539 Chemistry, Other *(see also BIOLOGICAL/Biochemistry)*

GEOLOGICAL & EARTH SCIENCES

542 Geochemistry
540 Geology
552 Geomorphology & Glacial Geology
544 Geophysics & Seismology
548 Mineralogy & Petrology
546 Paleontology
550 Stratigraphy & Sedimentation
558 Geological & Earth Sciences, General
559 Geological & Earth Sciences, Other

OCEAN/MARINE SCIENCES

585 Hydrology & Water Resources
590 Oceanography, Chemical & Physical
595 Marine Sciences
599 Ocean/Marine, Other

PHYSICS

560 Acoustics
576 Applied Physics
561 Atomic/Molecular/Chemical Physics
565 Biophysics *(also in BIOLOGICAL SCIENCES)*
574 Condensed Matter/Low Temperature Physics
568 Nuclear Physics
569 Optics/Photonics
564 Particle (Elementary) Physics
570 Plasma/Fusion Physics
572 Polymer Physics
578 Physics, General
579 Physics, Other

PSYCHOLOGY

600 Clinical Psychology
603 Cognitive Psychology & Psycholinguistics
606 Comparative Psychology
609 Counseling
612 Developmental & Child Psychology
618 Educational Psychology *(also in EDUCATION)*
615 Experimental Psychology
620 Family Psychology
613 Human Development & Family Studies
621 Industrial & Organizational *(see also BUSINESS MANAGEMENT/Organization Behavior)*
624 Personality Psychology
627 Physiological/Psychobiology Psychology
633 Psychometrics & Quantitative Psychology
636 School Psychology *(also in EDUCATION)*
639 Social Psychology
648 Psychology, General
649 Psychology, Other

SOCIAL SCIENCES

650 Anthropology
652 Area/Ethnic/Cultural/Gender Studies
657 Criminal Justice & Corrections
658 Criminology
662 Demography/Population Studies
668 Econometrics
667 Economics
670 Geography
674 International Relations/Affairs
676 Linguistics
678 Political Science & Government
682 Public Policy Analysis
686 Sociology
690 Statistics *(also in MATHEMATICS)*
694 Urban Affairs/Studies
695 Urban/City, Community & Regional Planning
698 Social Sciences, General
699 Social Sciences, Other

FIELDS NOT ELSEWHERE CLASSIFIED (NEC)

960 Architecture/Environmental Design
964 Family/Consumer Science/Human Science *(also in EDUCATION)*
968 Law
972 Library Science
974 Parks/Sports/Rec./Leisure/Fitness
976 Public Administration
980 Social Work
984 Theology/Religious Education *(see also OTHER HUMANITIES/Religion/Religious Studies)*
989 Other Fields, NEC

FIELD UNKNOWN

999 Unknown Field

Appendix C

Data on Science and Engineering Doctorates

APPENDIX TABLE C-1 Doctorates Awarded, by Selected Fields of Study, 2000-2005

Field	2000	2001	2002	2003	2004	2005
All fields	41,361	40,651	39,953	40,740	42,117	43,354
Science and engineering	25,966	25,496	24,582	25,274	26,272	27,974
Science	20,643	19,988	19,505	19,995	20,497	21,570
Biological/agricultural sciences	6,890	6,668	6,699	6,753	6,984	7,406
Earth, atmospheric, and ocean sciences	694	660	689	683	686	713
Mathematics/computer sciences	1,910	1,832	1,726	1,859	2,024	2,339
Physical sciences	3,378	3,364	3,185	3,289	3,338	3,647
Psychology	3,616	3,385	3,197	3,273	3,327	3,327
Social sciences	4,155	4,079	4,009	4,138	4,138	4,138
Engineering	5,323	5,508	5,077	5,279	5,775	6,404
Non-science and engineering	15,395	15,155	15,371	15,466	15,845	15,380

Source: National Science Foundation/Division of Science Resources Statistics, Survey of Earned Doctorates. Adapted from NSF (2006d): Table 1.

APPENDIX TABLE C-2 Doctorates Awarded to Women, by Selected Fields of Study, 2000-2005.

Field	2000	2001	2002	2003	2004	2005	2005(%)
All fields	18,126	17,855	18,117	18,496	19,157	19,564	45.1
Science and engineering	9,393	9,286	9,163	9,517	9,856	10,533	37.7
Science	8,555	8,356	8,272	8,606	8,835	9,359	43.4
Biological/agricultural sciences	2,939	2,889	2,864	2,978	3,155	3,481	47.0
Earth, atmospheric, and ocean sciences	196	198	211	213	238	243	34.1
Mathematics/computer sciences	400	431	432	440	504	551	23.6
Physical sciences	827	828	847	891	865	972	26.7
Psychology	2,410	2,260	2,132	2,231	2,246	2,264	68.0
Social sciences	1,783	1,750	1,786	1,853	1,827	1,848	44.7
Engineering	838	930	891	911	1,021	1,174	18.3
Non-science and engineering	8,733	8,569	8,954	8,979	9,301	9,031	58.7

Source: National Science Foundation/Division of Science Resources Statistics, Survey of Earned Doctorates. Adapted from NSF (2006d): Table 2.

APPENDIX TABLE C-3 Doctorates Awarded, by Selected Fields of Study and Citizenship, 2000–2005

Characteristic and field	2000	2001	2002	2003	2004	2005
All fields	41,361	40,651	39,953	40,740	42,117	43,354
Science and engineering	25,966	25,496	24,582	25,274	26,272	27,974
Science	20,643	19,988	19,505	19,995	20,497	21,570
Agricultural sciences	1,037	975	1,009	1,060	1,045	1,038
Biological sciences	5,853	5,693	5,690	5,693	5,939	6,368
Computer sciences	860	825	807	866	948	1,136
Earth, atmospheric, and ocean sciences	694	660	689	683	686	713
Mathematics	1,050	1,007	919	993	1,076	1,203
Physical sciences	3,378	3,364	3,185	3,289	3,338	3,647
Psychology	3,616	3,385	3,197	3,273	3,327	3,327
Social sciences	4,155	4,079	4,009	4,138	4,138	4,138
Engineering	5,323	5,508	5,077	5,279	5,775	6,404
Non-science and engineering	15,395	15,155	15,371	15,466	15,845	15,380
U.S. citizen or permanent resident, all fields	29,936	28,800	27,650	28,129	28,004	27,912
Science and engineering	17,116	16,319	15,511	15,733	15,744	16,024
Science	14,543	13,867	13,346	13,555	13,557	13,740
Agricultural sciences	559	506	492	562	566	557
Biological sciences	4,268	4,248	4,113	4,059	4,196	4,396
Computer sciences	458	424	420	446	447	473
Earth, atmospheric, and ocean sciences	474	411	433	451	438	442
Mathematics	574	526	443	517	511	541
Physical sciences	2,072	2,035	1,923	1,951	1,858	1,900
Psychology	3,230	2,977	2,793	2,855	2,788	2,891
Social sciences	2,908	2,740	2,729	2,714	2,753	2,540
Engineering	2,573	2,452	2,165	2,178	2,187	2,284
Non-science and engineering	12,820	12,481	12,139	12,396	12,260	11,888
U.S. citizen, all fields	27,986	26,975	25,998	26,499	26,466	26,312
Science and engineering	15,707	15,049	14,341	14,635	14,741	14,912
Science	13,484	12,896	12,448	12,723	12,796	12,913
Agricultural sciences	498	469	460	521	531	527
Biological sciences	3,904	3,909	3,798	3,796	3,963	4,141
Computer sciences	389	369	356	389	397	405
Earth, atmospheric, and ocean sciences	444	382	397	419	415	421
Mathematics	518	471	412	471	456	480
Physical sciences	1,871	1,854	1,762	1,800	1,740	1,768
Psychology	3,155	2,902	2,722	2,786	2,723	2,811
Social sciences	2,705	2,540	2,541	2,541	2,571	2,360
Engineering	2,223	2,153	1,893	1,912	1,945	1,999
Non-science and engineering	12,279	11,926	11,657	11,864	11,725	11,400
Non-U.S. citizen with permanent visa, all fields	1,950	1,825	1,652	1,630	1,538	1,600
Science and engineering	1,409	1,270	1,170	1,098	1,003	1,112

Science	1,059	971	898	832	761	827
Agricultural sciences	61	37	32	41	35	30
Biological sciences	364	339	315	263	233	255
Computer sciences	69	55	64	57	50	68
Earth, atmospheric, and ocean sciences	30	29	36	32	23	21
Mathematics	56	55	31	46	55	61
Physical sciences	201	181	161	151	118	132
Psychology	75	75	71	69	65	80
Social sciences	203	200	188	173	182	180
Engineering	350	299	272	266	242	285
Non-science and engineering	541	555	482	532	535	488
Non-U.S. citizen with temporary visa, all fields	9,660	9,800	9,731	10,589	11,617	12,824
Science and engineering	7,658	7,943	7,691	8,382	9,151	10,404
Science	5,207	5,156	5,042	5,472	5,843	6,650
Agricultural sciences	443	400	433	427	418	415
Biological sciences	1,385	1,242	1,292	1,398	1,470	1,677
Computer sciences	363	358	348	378	459	599
Earth, atmospheric, and ocean sciences	182	219	223	201	224	233
Mathematics	443	434	440	440	528	602
Physical sciences	1,148	1,205	1,106	1,216	1,346	1,550
Psychology	164	152	156	195	188	210
Social sciences	1,079	1,146	1,044	1,217	1,210	1,364
Engineering	2,451	2,787	2,649	2,910	3,308	3,754
Non-science and engineering	2,002	1,857	2,040	2,207	2,466	2,420
Non-U.S. citizen, all fields	11,610	11,625	11,383	12,219	13,155	14,424
Science and engineering	9,067	9,213	8,861	9,480	10,154	11,516
Science	6,266	6,127	5,940	6,304	6,604	7,477
Agricultural sciences	504	437	465	468	453	445
Biological sciences	1,749	1,581	1,607	1,661	1,703	1,932
Computer sciences	432	413	412	435	509	667
Earth, atmospheric, and ocean sciences	212	248	259	233	247	254
Mathematics	499	489	471	486	583	663
Physical sciences	1,349	1,386	1,267	1,367	1,464	1,682
Psychology	239	227	227	264	253	290
Social sciences	1,282	1,346	1,232	1,390	1,392	1,544
Engineering	2,801	3,086	2,921	3,176	3,550	4,039
Non-science and engineering	2,543	2,412	2,522	2,739	3,001	2,908
Citizenship unknown, all fields	1,765	2,051	2,572	2,022	2,496	2,618
Science and engineering	1,192	1,234	1,380	1,159	1,377	1,546
Science	893	965	1,117	968	1,097	1,180
Agricultural sciences	35	69	84	71	61	66
Biological sciences	200	203	285	236	273	295
Computer sciences	39	43	39	42	42	64
Earth, atmospheric, and ocean sciences	38	30	33	31	24	38
Mathematics	33	47	36	36	37	60
Physical sciences	158	124	156	122	134	197
Psychology	222	256	248	223	351	226

Social sciences	168	193	236	207	175	234
Engineering	299	269	263	191	280	366
Non-science and engineering	573	817	1,192	863	1,119	1,072

Source: National Science Foundation/Division of Science Resources Statistics, Survey of Earned Doctorates. Adapted from NSF (2006d): Table 3.

APPENDIX TABLE C-4 Doctorates Awarded to U.S. Citizens, by Race/Ethnicity and Major Field of Study of Recipients, 2000-2005.

Race/ethnicity and field	2000	2001	2002	2003	2004	2005
All fields	27,986	26,975	25,998	26,499	26,466	26,312
Science and engineering	15,707	15,049	14,341	14,635	14,741	14,912
Science	13,484	12,896	12,448	12,723	12,796	12,913
Agricultural sciences	498	469	460	521	531	527
Biological sciences	3,904	3,909	3,798	3,796	3,963	4,141
Computer sciences	389	369	356	389	397	405
Earth, atmospheric, and ocean sciences	444	382	397	419	415	421
Mathematics	518	471	412	471	456	480
Physical sciences	1,871	1,854	1,762	1,800	1,740	1,768
Psychology	3,155	2,902	2,722	2,786	2,723	2,811
Social sciences	2,705	2,540	2,541	2,541	2,571	2,360
Engineering	2,223	2,153	1,893	1,912	1,945	1,999
Non-science and engineering	12,279	11,926	11,657	11,864	11,725	11,400
American Indian/Alaska Native, all fields	169	148	146	134	129	139
Science and engineering	88	71	66	72	59	66
Science	80	65	60	61	54	58
Agricultural sciences	5	0	2	4	4	3
Biological sciences	17	15	12	11	14	12
Computer sciences	1	0	1	2	2	1
Earth, atmospheric, and ocean sciences	4	0	0	2	4	4
Mathematics	2	1	3	2	0	0
Physical sciences	9	12	7	2	5	4
Psychology	22	17	15	22	12	15
Social sciences	20	20	20	16	13	19
Engineering	8	6	6	11	5	8
Non-science and engineering	81	77	80	62	70	73
Asian, all fields[a]	1,365	1,411	1,364	1,372	1,451	1,493
Science and engineering	992	1,053	1,035	1,008	1,066	1,114
Science	751	790	788	797	838	872
Agricultural sciences	15	11	13	9	13	13
Biological sciences	318	361	373	379	380	409
Computer sciences	41	36	50	45	42	52
Earth, atmospheric, and ocean sciences	8	12	5	8	6	11
Mathematics	44	32	19	38	29	38
Physical sciences	101	123	118	110	126	140
Psychology	121	100	105	104	134	110
Social sciences	103	115	105	104	108	99
Engineering	241	263	247	211	228	242
Non-science and engineering	373	358	329	364	385	379
Black/African American, all fields	1,631	1,611	1,665	1,708	1,881	1,688
Science and engineering	646	630	636	615	689	640

Science	572	548	560	544	605	555
Agricultural sciences	12	9	9	18	18	13
Biological sciences	110	125	114	100	136	142
Computer sciences	13	13	17	17	15	15
Earth, atmospheric, and ocean sciences	3	5	4	11	6	6
Mathematics	14	17	13	14	8	16
Physical sciences	52	46	62	48	49	47
Psychology	184	167	162	163	204	159
Social sciences	184	166	179	173	169	157
Engineering	74	82	76	71	84	85
Non-science and engineering	985	981	1,029	1,093	1,192	1,048
Hispanic, all fields[b]	1,182	1,122	1,237	1,280	1,178	1,294
Science and engineering	651	581	652	659	645	722
Science	582	507	564	568	572	649
Agricultural sciences	20	21	20	19	15	14
Biological sciences	155	149	161	158	175	207
Computer sciences	11	6	13	9	12	9
Earth, atmospheric, and ocean sciences	16	5	11	10	10	15
Mathematics	13	14	11	15	21	23
Physical sciences	71	61	55	61	56	69
Psychology	203	155	179	162	164	178
Social sciences	93	96	114	134	119	134
Engineering	69	74	88	91	73	73
Non-science and engineering	531	541	585	621	533	572
Mexican American, all fields	400	379	400	458	431	497
Science and engineering	206	170	183	209	217	237
Science	180	153	167	194	195	213
Agricultural sciences	6	4	7	7	6	3
Biological sciences	44	47	52	53	54	68
Computer sciences	3	2	3	3	3	2
Earth, atmospheric, and ocean sciences	4	2	4	0	3	3
Mathematics	8	6	2	4	7	7
Physical sciences	22	15	15	25	20	19
Psychology	60	49	50	51	51	58
Social sciences	33	28	34	51	51	53
Engineering	26	17	16	15	22	24
Non-science and engineering	194	209	217	249	214	260
Puerto Rican, all fields	326	296	345	260	259	264
Science and engineering	183	149	196	147	140	159
Science	163	136	170	128	121	143
Agricultural sciences	6	6	9	3	4	5
Biological sciences	33	48	56	34	51	53
Computer sciences	4	1	4	2	4	0
Earth, atmospheric, and ocean sciences	5	1	3	4	5	5
Mathematics	2	2	3	3	2	5
Physical sciences	20	14	20	17	16	17
Psychology	74	39	50	47	19	37

Social sciences	19	25	25	18	20	21
Engineering	20	13	26	19	19	16
Non-science and engineering	143	147	149	113	119	105
Other Hispanic, all fields	456	447	492	562	488	533
Science and engineering	262	262	273	303	288	326
Science	239	218	227	246	256	293
Agricultural sciences	8	11	4	9	5	6
Biological sciences	78	54	53	71	70	86
Computer sciences	4	3	6	4	5	7
Earth, atmospheric, and ocean sciences	7	2	4	6	2	7
Mathematics	3	6	6	8	12	11
Physical sciences	29	32	20	19	20	33
Psychology	69	67	79	64	94	83
Social sciences	41	43	55	65	48	60
Engineering	23	44	46	57	32	33
Non-science and engineering	194	185	219	259	200	207
White, all fields	22,970	21,869	20,757	20,872	20,762	20,845
Science and engineering	12,921	12,225	11,486	11,612	11,630	11,848
Science	11,156	10,575	10,091	10,155	10,168	10,327
Agricultural sciences	433	417	407	452	457	473
Biological sciences	3,195	3,143	3,028	2,978	3,104	3,248
Computer sciences	311	300	267	294	308	305
Earth, atmospheric, and ocean sciences	398	345	361	369	364	374
Mathematics	437	396	353	379	376	387
Physical sciences	1,594	1,547	1,464	1,481	1,410	1,442
Psychology	2,569	2,372	2,188	2,222	2,105	2,243
Social sciences	2,219	2,055	2,023	1,980	2,044	1,855
Engineering	1,765	1,650	1,395	1,457	1,462	1,521
Non-science and engineering	10,049	9,644	9,271	9,260	9,132	8,997
Education	4,329	4,158	4,007	4,103	3,978	3,928
Health	940	885	944	934	938	929
Humanities	3,583	3,472	3,264	3,229	3,166	3,020
Professional/other/unknown	1,197	1,129	1,056	994	1,050	1,120
Other/unknown, all fields[c]	669	814	829	1,133	1,065	853
Science and engineering	409	489	466	669	652	522
Science	343	411	385	598	559	452
Agricultural sciences	13	11	9	19	24	11
Biological sciences	109	116	110	170	154	123
Computer sciences	12	14	8	22	18	23
Earth, atmospheric, and ocean sciences	15	15	16	19	25	11
Mathematics	8	11	13	23	22	16
Physical sciences	44	65	56	98	94	66
Psychology	56	91	73	113	104	106
Social sciences	86	88	100	134	118	96
Engineering	66	78	81	71	93	70
Non-science and engineering	260	325	363	464	413	331

[a] Pacific Islanders are included in this category prior to 2001.

[b] Includes Mexican American, Puerto Rican, and other Hispanic.
[c] Native Hawaiian, other Pacific Islanders, and multiple race/ethnicity are included in this category from 2001 forward.
Note: Categories are grouped differently from questionnaire and summary reports in that linguistics, history of science, American studies, and archaeology are included in social sciences and not in humanities, and public administration is included in social sciences.
Source: National Science Foundation/Division of Science Resources Statistics, Survey of Earned Doctorates. Adapted from NSF (2006d): Table 3.

Appendix D
APPLICATION FOR RESEARCH ASSOCIATE PROGRAMS

OFFICE USE ONLY	
ID#	
Postdoctoral	
Senior	

APPLICATION

THIS IS AN EXAMPLE-ONLY APPLICATION – DO NOT SIGN. DO NOT SUBMIT.

Applicant Last or Family Name	*First Name*
Middle Name	*Maiden Name* **(if applicable)**

CURRENT Address	**Home or Institution, MailCode/Stop, Bldg./Room, Number/Street**	*City* *State / Province*	*Zip* **(Postal)** *Code* *Country*
PERMANENT Address	**Home or Institution, MailCode/Stop, Bldg./Room, Number/Street**	*City* *State / Province*	*Zip* **(Postal)** *Code* *Country*
CITIZENSHIP	**Indicate ALL countries of which you are a citizen.**	*Passport Expiration Date(s)*	

If you are a naturalized US citizen, enter your naturalization date and number.	*Date* **Month / Day / Year**	*Number*
If you are a non-US citizen already in the US, enter the type of visa you hold.	*Visa Type and Category*	*Date J-1 Status* **(DS-2019)** *Expires*
If you are a US legal permanent resident, enter your alien registration number and *enclose* ***a copy of your alien registration (green) card.***	*Alien Registration Number*	

EDUCATION – List in order, beginning with the most recent degree awarded or expected.

Complete Name of University or College **including City, State/Province, Country**	*Inclusive Dates* **Year to Year**	*Degree*	*Awarded or Expected* **Month / Year**	*Degree Discipline / Field Code* **refer to Field Reference List**	
	-				
	-				
	-				
	-				
	-				
	-				

All transcripts for Postdoctoral applicants must be enclosed with the application package.

HONORS AND AWARDS

Title	*Complete Name of Institution* **including City, State/Province, Country**	*Inclusive Dates* **Year to Year**
		-
		-
		-

Title of Research Proposal

APPLICATION
continued

OFFICE USE ONLY	
Last Name	**ID**

Have you previously applied for a National Academies Associateship? **NO** **YES** (*Agency or Agencies*)	*Year*
Are you a former National Academies Research Associate? **NO** **YES** (*Agency or Agencies*)	*Tenure Dates* **Year to Year** -

EMPLOYMENT – Professional, Scientific, Administrative, etc. List in order, beginning with most recent.

Name of Organization **including City, State/Province, Country**	*Employment Sector*	*Title or Academic Rank*	*Inclusive Dates* **Year to Year**
			-
			-
			-
			-

Will you be on official leave from your current position, to which you will return?

NO **YES**

PROGRAM INFORMATION – You may apply for a maximum of three (3) different Agencies.

Agency or Agencies	*Laboratory* or *NASA Center*	*Proposed Research Adviser*	*Research Opportunity Number*

Field of Proposed Research. Field Name		*Code* Code
Proposed Length of Tenure **(for Seniors only)** **months**	***Proposed Starting Date*** **(Month / Day / Year)**	

REFERENCES – Professional, Scientific, Administrative, etc.

Enter the names, titles, and professional addresses of four (4) respondents who are familiar with your research. For recent doctoral recipients, the first name listed should be that of the thesis adviser.

Full Name of Respondent	*Complete Professional Mailing Address of Respondent*
1)	
2)	
3)	
4)	
extra 5)	

APPLICATION
continued

OFFICE USE ONLY

ID#

This information is used by the NRC and sponsors to process awards. Optional information on race and ethnicity is for statistical purposes. Information on this page is not seen by reviewers.

APPLICANT

Applicant Last or Family Name			*First Name*	
Date of Birth **Month / Day / Year**			*Place of Birth* **City, State/Province, Country**	
U.S. Social Security Number	*Sex* **Male Female**	*Marital Status* **Single Married**	*Race*	*Ethnicity*

CONTACT INFORMATION

Office Phone	*Home Phone*
Fax	*E-mail*

SPOUSE

Spouse's Last or Family Name	*First Name*
Date of Birth **Month / Day / Year**	*Place of Birth* **City, State/Province, Country**

Dependent Full Name	*Date of Birth* **Month / Day / Year**	*Place of Birth* **City, State/Province, Country**

APPLICATION
continued

OFFICE USE ONLY

Last Name	**ID#**

To assist us in making information available to a greater number of potential applicants, it is important for us to learn how you initially heard about the National Academies Research Associateship Programs.

Please check ONLY ONE of the following:

colleague or fellow graduate student

Ph.D. thesis adviser or other professor

university placement office

former or current NRC Research Associate

Research Adviser or other scientific staff at the federal Laboratory

Research Associateship Programs' staff member at professional scientific meeting

Title of Meeting	*Date of Meeting* **Month / Year**

Advertisement in professional publication

Name of Publication

Other

Please Specify

To which review are you applying?

March Review (deadline February 1)
June Review (deadline May 1)
September Review (deadline August 1)
January Review (deadline November 1)

THE NATIONAL ACADEMIES
Advisers to the Nation on Science, Engineering, and Medicine

Research Associateship Programs

PREVIOUS AND CURRENT RESEARCH

to be completed by Postdoctoral applicants only

OFFICE USE ONLY

ID#

Applicant Last or Family Name		*First Name*
Middle Name		*Maiden Name* **(if applicable)**
Date of Ph.D. **Month/Year** /	*Complete Name of University or College*	*Thesis Adviser*

Title of Ph.D. Dissertation

Status of Ph.D. Dissertation

Published Accepted for publication In preparation for publication Not to be published

Attach a concise description of all investigations, stating where, when, and with whom they were carried out.
(Maximum of 1200 words, double-spaced, 12-point font. DO NOT SEND REPRINTS.)

Attach a list of publications in the following order: 1) refereed journal articles; 2) books; 3) published proceedings; 4) non-refereed articles; and, 5) patents. Citations should include the following: a) authors; b) year of publication; c) title; d) full name of journal; e) volume number; and f) page number(s).
(Maximum of 1800 words, double-spaced, 12-point font. DO NOT SEND REPRINTS.)

THE NATIONAL ACADEMIES
Advisers to the Nation on Science, Engineering, and Medicine

Research Associateship Programs

PREVIOUS AND CURRENT RESEARCH

to be completed by senior applicants only

OFFICE USE ONLY

ID#

Applicant Last or Family Name	*First Name*
Middle Name	*Maiden Name* **(if applicable)**

Attach a concise description of all investigations, stating where, when, and with whom they were carried out.

(Maximum of 1200 words, double-spaced, 12-point font. DO NOT SEND REPRINTS.)

Attach a list of publications within the past five (5) years in the following order: 1) refereed journal articles; 2) books; 3) published proceedings; 4) non-refereed articles; and, 5) patents. Citations should include the following: a) authors; b) year of publication; c) title; d) full name of journal; e) volume number; and f) page number(s).

(Maximum of 1800 words, double-spaced, 12-point font. DO NOT SEND REPRINTS.)

THE NATIONAL ACADEMIES

Advisers to the Nation on Science, Engineering, and Medicine

Research Associateship Programs

OFFICE USE ONLY

ID#

REFERENCE REPORT

APPLICANT: attach a brief abstract of your Research Proposal.
If you are applying to more than one Agency with different Research Proposals, please submit separate abstracts, and identify the Agency to which each refers. Maximum of 350 words per abstract.

Applicant Last or Family Name	*First Name*	
Field of Proposed Research	*Agency or Agencies*	*Laboratory* or *NASA Center*
Title of Research Proposal		

RESPONDENT: return the Reference <u>to the applicant</u> --
completed, signed, in a sealed envelope with respondent's signature across the envelope flap.

Full Name & Title of Respondent	*Institutional Affiliation*	
Address	*Office Phone*	*E-mail*

1) I have known this applicant in the following capacities (you may check more than one)

as an undergraduate
as a graduate student
as a teaching/research assistant
as my thesis advisee
as a professional colleague
by reputation only

2) I was acquainted with the professional work of this applicant from ______ to ______
Month / Year to **Month / Year**

3) I had a **poor** **fair** **good** **excellent** opportunity to observe the quality of this applicant's work.

4) If the applicant is/was a student, how does he/she compare with students currently in your department?

Lower half **Upper half** **Top 25%** **Top 10%** **Top 1%**

5) Please indicate on this scale, your overall impression of this applicant. **(Check ONLY one.)**

Below Average	Average	Above Agerage	Excellent	Outstanding	Inadequate Opportunity to Observe

REFERENCE REPORT

continued

OFFICE USE ONLY

ID#

Applicant Last or Family Name	*First Name*

6) Please comment on the Applicant's scientific and technical abilities, both in comparison with other scientists and engineers with similar training and experience and with respect to the proposed research (see attached Abstract). Include in your assessment the following: a) knowledge of the field; b) skill in experimental design; c) technical abilities; d) innovative abilities; e) ability to work independently; f) analytical abilities; and, g) skills in interpreting and reporting research.

REFERENCE REPORT

continued

OFFICE USE ONLY

ID#

Applicant Last or Family Name	*First Name*

RESPONDENT: Your response below is necessary if applicant requests information from the file.

I ask that the National Academies maintain the confidentiality of my identity to the extent permitted by law.
I further ask the National Academies to maintain the confidentiality of these comments to the extent permitted by law.

I ask that the National Academies maintain the confidentiality of my identity to the extent permitted by law.
I do not ask the National Academies to hold my comments in confidence.

My preparation of this Reference Report is not conditioned on the request that the National Academies hold my identity or comments in confidence.

Signature of Respondent Date

Please also print your name

Respondent for a Senior Applicant may write a Letter of Reference instead of completing the Reference Report form, but should also address the points listed on this form.

The Reference Report or Letter of Reference must be in English, must have a current date, and IF sending a hard copy, must bear the *original* signature (not photocopy, fax, or electronic) **of the respondent who is listed on the Application.**

Respondent *may* return the Reference Report or Letter of Reference to the applicant -- completed, signed, and in a sealed envelope with the respondent's signature clearly written across the envelope flap.

OR

Respondent *may* send the completed Reference Report or Letter of Reference directly to the Associateship Programs office (rap@nas.edu) as an e-mail attachment. It must come directly from the Respondent so we can accept the name on the 'From' line as the official signature.

THE NATIONAL ACADEMIES
Advisers to the Nation on Science, Engineering, and Medicine

Research Associateship Programs

OFFICE USE ONLY

ID#

RESEARCH PROPOSAL

Applicant Last or Family Name	*First Name*
Middle Name	*Maiden Name* **(if applicable)**

PROPOSED RESEARCH ADVISER INFORMATION

Proposed Research Adviser	*Agency or* ***Agencies***	*Laboratory* or *NASA Center*
1)		
2)		
3)		

Title of Research Proposal

ATTACH DETAILED RESEARCH PROPOSAL

(Maximum of 3000 words, double-spaced, 12-point font)

The *Research Proposal* should be sufficiently complete for outside peer review purposes. Description of the proposed research must include the following: a) statement of problem; b) background and relevance to previous work; c) general methodology and procedure to be followed; d) explanation of new or unusual techniques; e) expected results and their significance and application; and, f) literature citations where appropriate.

APPENDIX

Research Proposal

OFFICE USE ONLY

ID#

Applicant Last or Family Name	*First Name*

ANTICIPATED RESEARCH NEEDS -- Indicate special requirements necessary to conduct your research.
(Entering information electronically expands the field to accommodate all of the space you need.)

1) *Describe activities related to the acquisition or collection of data, such as field activities, research voyages, or observatory use*

2) *Computer resources*

3) *Specialized equipment*

4) *Other*

THE NATIONAL ACADEMIES
Advisers to the Nation on Science, Engineering, and Medicine

Research Associateship Programs

OFFICE USE ONLY

ID#

LABORATORY/CENTER REVIEW

THE PROPOSED RESEARCH ADVISER COMPLETES THIS SECTION.

Applicant Last or Family Name	*First Name*
Middle Name	*Maiden Name* **(if applicable)**

Agency or Agencies	*Laboratory* or *NASA Center*	*Research Opportunity Number*
Title of Research Proposal		

Proposed Length of Tenure **for seniors only** **number of months**	*Proposed Starting Date* **Month / Day / Year**	*Level* Postdoctoral Senior

PROPOSED RESEARCH ADVISER INFORMATION

Adviser Name	*Adviser Office Phone*	*Adviser E-mail*	
Adviser Address	*City*	*State*	*Zip* **(Postal)** *Code*

Please address the overall scientific quality of the research proposed by this applicant, including the specific points indicated on the following page. Be sure the applicant's name is at the top of each page.

Recommended for review

Not recommended for review – no Laboratory interest.

Signature of Proposed Research Adviser
Please also print your name

Date

After completing the above portion, sign, date and forward this form to the Laboratory or Center Program Representative.

LABORATORY/CENTER PROGRAM REPRESENTATIVE'S RECOMMENDATION

The Laboratory/Center **recommends** this Research Proposal for review.

The Laboratory/Center **does not recommend** this Research Proposal for review.

Laboratory/Center Program Representative's Comments

Signature of Laboratory/Center Program Representative

Date

Document should be sent by express delivery to:

Research Associateship Programs
The National Academies
2001 Wisconsin Avenue, NW [GR 322A]
Washington, DC 20007

LABORATORY/CENTER REVIEW

OFFICE USE ONLY

ID#

Applicant Last or Family Name	*First Name*

1) Are the proposed objectives realistic? 2) Does the proposal reflect innovative thinking? 3) Is the technical work plan sound, and does it incorporate state-of-the-art methods? 4) Can the research be accomplished in the proposed timeframe? Please also comment on the relevance of the proposed research to the mission of your agency. If specialized equipment or facilities are needed for the proposed research, please address the availability of these. If animal or human subjects will be used in the proposed research, indicate if an IACUC or IRB approval has been or will be obtained.

Appendix E
EXAMINATION OF APPLICATIONS TO PRESOPECTIVE PROGRAMS

In the next table, we examine the number of applicants to NIST and non-NIST postdoc positions and the number of those accepted to the respective programs. The proportion of applicants and awardees is contrasted with the relevant pool of potential applicants. In the case of applicants to NIST postdocs, the pool is the sum of U.S. citizens, who received a Ph.D. in the five years prior to the application year, including the application year, in science and engineering.[48] We also look at the subset of Ph.D.s who were U.S. citizens and who intended to pursue a postdoctoral appointment. (These are doctorates who had definite plans for postdoctoral study or research.) The figure presented in a five-year moving average of those with definite plans.

In the case of applicants to non-NIST postdocs, the pool consists of all Ph.D.s in the five years prior to the application year, including the application year, in science and engineering.[49] Again, we also look at the subset of Ph.D.s who intended to pursue a postdoctoral appointment. Data on Ph.D.s and those intending to do postdocs is taken from the National Science Foundation's Survey of Earned Doctorates, described in Chapter 1. (Selected pages from the 2007 survey questionnaire is found in Appendix B.) For comparative purposes, we focus on data from 1965 through 2005, which is the most current SED data available at this time.

---

[48] Science and engineering is defined as those fields on the SED associated with 000 to 599 field codes. In 16 cases the Ph.D. field was unknown. These cases were dropped from the analysis. This leaves a total of 667,255 individuals with S&E degrees.

[49] This is a less-precise comparison group, as some of the postdoc positions in the non-NIST group are open to a subset, e.g., U.S. citizens or permanent residents, while others are open to U.S. citizens and foreign nationals.

APPENDIX TABLE E-1 Doctoral Pool of Potential Postdocs, Applications, and Acceptances, by Research Associateship Program

A. NIST

Year	Ph.D.s	Intending to be Postdoc	%	Applications	Accepted	%
1965	26225	4607	18	64	61	95
1966	29173	5312	18	70	36	51
1967	32493	5979	18	93	53	57
1968	36198	6686	18	108	15	14
1969	40364	7572	19	190	16	8
1970	44731	8654	19	214	15	7
1971	49026	9818	20	252	16	6
1972	51840	10815	21	238	19	8
1973	53331	11497	22	153	20	13
1974	52574	11535	22	140	17	12
1975	50922	11563	23	144	17	12
1976	48563	11513	24	129	23	18
1977	46548	11448	25	111	22	20
1978	44691	11643	26	104	23	22
1979	44340	12307	28	105	24	23
1980	43772	12621	29	81	25	31
1981	43562	12780	29	87	22	25
1982	43745	13000	30	90	16	18
1983	43998	13165	30	131	24	18
1984	44163	13387	30	108	26	24
1985	44213	13487	31	102	28	27
1986	44170	13738	31	48	N/A	N/A
1987	44147	14087	32	122	22	18
1988	44813	14595	33	128	22	17
1989	45502	14869	33	105	20	19
1990	46683	15309	33	134	25	19
1991	48316	15691	32	200	26	13
1992	49781	16037	32	226	31	14
1993	50895	16273	32	237	34	14
1994	52061	16582	32	254	40	16
1995	53121	16741	32	222	38	17
1996	53806	16775	31	156	57	37
1997	54973	16759	30	170	40	24
1998	56091	16688	30	120	58	48
1999	56379	16509	29	132	39	30
2000	56359	16513	29	118	57	48
2001	56016	16509	29	79	35	44
2002	54725	16536	30	143	52	36
2003	53577	16644	31	165	56	34
2004	53165	17141	32	166	48	29
2005	53129	17662	33	218	59	27
2006	N/A	N/A	N/A	183	59	32
2007	N/A	N/A	N/A	107	47	44
Total				6147	1383	22

B. Other

Year	Ph.D.s	Intending to be Postdoc	%	Applications	Accepted	%
1965	32522	5651	17	132	113	86
1966	36488	6519	18	191	135	71
1967	40892	7421	18	202	147	73
1968	45757	8366	18	350	179	51
1969	51091	9504	19	603	137	23
1970	56538	10860	19	795	172	22
1971	62029	12308	20	1131	216	19
1972	66074	13606	21	1317	253	19
1973	68702	14580	21	1187	202	17
1974	69315	14804	21	1214	205	17
1975	68445	14937	22	1165	228	20
1976	66309	14949	23	988	192	19
1977	64129	14799	23	776	140	18
1978	61943	14886	24	827	176	21
1979	60868	15494	25	779	180	23
1980	60034	15780	26	699	195	28
1981	59974	15841	26	753	221	29
1982	60675	16064	26	788	195	25
1983	61777	16270	26	732	196	27
1984	62887	16623	26	826	220	27
1985	64313	16947	26	827	239	29
1986	65918	17733	27	867	284	33
1987	67767	18752	28	907	260	29
1988	70666	20038	28	862	310	36
1989	73884	21135	29	881	315	36
1990	77765	22448	29	849	286	34
1991	82275	23711	29	1079	318	29
1992	86834	24964	29	1024	291	28
1993	90765	25898	29	1288	355	28
1994	94805	26972	28	1179	369	31
1995	98180	27790	28	863	275	32
1996	100994	28463	28	739	293	40
1997	103057	28601	28	743	283	38
1998	104670	28817	28	580	220	38
1999	104048	28629	28	523	207	40
2000	103244	28469	28	502	222	44
2001	101736	28046	28	547	293	54
2002	99768	28179	28	730	398	55
2003	98181	28656	29	808	308	38
2004	99059	29869	30	712	216	30
2005	101539	31487	31	755	170	23
2006	N/A	N/A	N/A	428	146	34
2007	N/A	N/A	N/A	150	50	33
Total				33298	9810	29

Source: Doctoral pool data from NSF, Survey of Earned Doctorates; other data from NRC DataRAP database.

Next, we disaggregated the total pool of potential applicants, applications, and acceptances by discipline. Perhaps, reflecting the changing nature of disciplines over the years, and the fact that applicants are supposed to identify a field for all degrees received, there are a large number of fields that applicants have put over the years. There were 1398 choices applicants could pick. Several are a bit confusing. As we are interested in doctoral degrees, applicants picked 536 different fields for their Ph.D.s. To simplify matters, we created a small group of major categories: agricultural sciences and natural resources; biological, biomedical, and health sciences; engineering; mathematical and computer sciences; and the physical sciences to examine. (A list of which fine fields are part of each of these major categories is found in Appendix F. There was some amount of subjectivity in this process, particularly in terms of whether a field should be included in these categories or not, but it should not effect the general results presented in the following tables.) We contrast these data to groups of Ph.D.s from the SED where agriculture and natural resources is anyone with a field code of 000 to 099; biological and health is anyone with a field code of 100-299; engineering is anyone with a field code of 300-399, mathematics and computer science is anyone with a field code of 400-499; and the physical science is anyone with a field code of 500-599.

APPENDIX TABLE E-2 Doctoral Pool of Potential Postdocs, Applications, and Acceptances in the Agricultural Sciences and Natural Resources, by Research Associateship Program

A. NIST

Year	Ph.D.s	Intending to be Postdoc	%	Applications	% of All	Accepted	%
1965	1780	133	7	0	0	0	N/A
1966	1817	130	7	1	1	0	0
1967	1877	136	7	0	0	0	N/A
1968	1955	125	6	0	0	0	N/A
1969	2131	145	7	0	0	0	N/A
1970	2386	152	6	0	0	0	N/A
1971	2706	175	6	0	0	0	N/A
1972	2943	187	6	0	0	0	N/A
1973	3153	224	7	0	0	0	N/A
1974	3185	230	7	0	0	0	N/A
1975	3209	242	8	0	0	0	N/A
1976	3107	251	8	0	0	0	N/A
1977	3041	248	8	0	0	0	N/A
1978	3016	232	8	0	0	0	N/A
1979	3060	241	8	0	0	0	N/A
1980	3071	259	8	0	0	0	N/A
1981	3164	268	8	1	1	0	0
1982	3335	316	9	0	0	0	N/A
1983	3440	348	10	0	0	0	N/A
1984	3523	392	11	0	0	0	N/A
1985	3623	416	11	0	0	0	N/A
1986	3597	454	13	0	0	N/A	N/A
1987	3510	479	14	0	0	0	N/A
1988	3401	513	15	0	0	0	N/A
1989	3389	543	16	0	0	0	N/A
1990	3321	579	17	0	0	0	N/A
1991	3294	595	18	0	0	0	N/A
1992	3210	596	19	0	0	0	N/A
1993	3093	577	19	0	0	0	N/A
1994	2976	559	19	0	0	0	N/A
1995	2846	521	18	0	0	0	N/A
1996	2746	490	18	0	0	0	N/A
1997	2708	462	17	0	0	0	N/A
1998	2737	457	17	0	0	0	N/A
1999	2670	434	16	0	0	0	N/A
2000	2602	418	16	0	0	0	N/A
2001	2523	418	17	0	0	0	N/A
2002	2415	421	17	0	0	0	N/A
2003	2367	430	18	0	0	0	N/A
2004	2435	476	20	0	0	0	N/A
2005	2498	510	20	0	0	0	N/A
2006	N/A	N/A	N/A	0	0	0	N/A
2007	N/A	N/A	N/A	0	0	0	N/A
Total				2	0	0	0

B. Other

Year	Ph.D.s	Intending to be Postdoc	%	Applications	% of All	Accepted	%
1965	2467	192	8	3	2	2	67
1966	2605	199	8	2	1	0	0
1967	2741	225	8	4	2	1	25
1968	2959	239	8	4	1	2	50
1969	3257	275	8	6	1	1	17
1970	3599	294	8	5	1	1	20
1971	4096	337	8	14	1	2	14
1972	4507	355	8	7	1	1	14
1973	4857	397	8	13	1	0	0
1974	5046	407	8	9	1	0	0
1975	5195	427	8	9	1	2	22
1976	5072	427	8	8	1	1	13
1977	4983	428	9	10	1	0	0
1978	4962	408	8	5	1	0	0
1979	4967	404	8	3	0	0	0
1980	4972	412	8	5	1	1	20
1981	5173	423	8	15	2	4	27
1982	5375	469	9	6	1	0	0
1983	5535	500	9	15	2	2	13
1984	5682	563	10	8	1	1	13
1985	5868	600	10	11	1	2	18
1986	5873	681	12	8	1	2	25
1987	5858	727	12	8	1	1	13
1988	5856	808	14	12	1	4	33
1989	5952	893	15	13	1	4	31
1990	6015	992	16	4	0	2	50
1991	6101	1036	17	6	1	3	50
1992	6190	1104	18	8	1	2	25
1993	6125	1103	18	11	1	5	45
1994	6113	1103	18	9	1	3	33
1995	6004	1068	18	14	2	6	43
1996	5968	1059	18	15	2	5	33
1997	5880	1028	17	23	3	9	39
1998	5969	1060	18	5	1	1	20
1999	5846	1061	18	8	2	3	38
2000	5719	1057	18	13	3	6	46
2001	5526	1063	19	7	1	3	43
2002	5426	1087	20	11	2	6	55
2003	5274	1100	21	7	1	4	57
2004	5312	1161	22	14	2	1	7
2005	5388	1215	23	10	1	3	30
2006	N/A	N/A	N/A	5	1	2	40
2007	N/A	N/A	N/A	4	3	0	0
Total				377	1	98	26

Source: Doctoral pool data from NSF, Survey of Earned Doctorates; other data from NRC DataRAP database.

APPENDIX TABLE E-3. Doctoral Pool of Potential Postdocs, Applications, and Acceptances in the Biological, Biomedical, and Health Sciences, by Research Associateship Program

A. NIST

Year	Ph.D.s	Intending to be Postdoc	%	Applications	% of All	Accepted	%
1965	6792	1945	29	0	0	N/A	N/A
1966	7560	2244	30	0	0	N/A	N/A
1967	8381	2542	30	0	0	N/A	N/A
1968	9512	3004	32	0	0	N/A	N/A
1969	10853	3464	32	1	1	1	100
1970	12329	4046	33	0	0	N/A	N/A
1971	13988	4632	33	2	1	0	0
1972	15294	5142	34	0	0	N/A	N/A
1973	16267	5399	33	0	0	N/A	N/A
1974	16584	5491	33	1	1	0	0
1975	16710	5647	34	0	0	N/A	N/A
1976	16605	5808	35	0	0	N/A	N/A
1977	16574	6042	36	2	2	0	0
1978	16558	6466	39	1	1	0	0
1979	17017	7093	42	2	2	2	100
1980	17495	7532	43	0	0	N/A	N/A
1981	17936	7919	44	0	0	N/A	N/A
1982	18482	8214	44	4	4	0	0
1983	18785	8382	45	2	2	0	0
1984	19074	8495	45	4	4	1	25
1985	19021	8484	45	2	2	0	0
1986	18865	8507	45	0	0	N/A	N/A
1987	18573	8493	46	1	1	0	0
1988	18641	8644	46	0	0	N/A	N/A
1989	18617	8674	47	0	0	N/A	N/A
1990	18831	8812	47	2	1	2	100
1991	19230	8879	46	2	1	0	0
1992	19781	9041	46	3	1	1	33
1993	20322	9138	45	0	0	N/A	N/A
1994	20856	9240	44	1	0	0	0
1995	21374	9235	43	5	2	0	0
1996	21762	9253	43	4	3	2	50
1997	22252	9238	42	2	1	0	0
1998	22770	9208	40	5	4	3	60
1999	23018	9144	40	4	3	2	50
2000	23609	9280	39	6	5	5	83
2001	24101	9395	39	0	0	N/A	N/A
2002	24395	9553	39	4	3	2	50
2003	24509	9639	39	5	3	2	40
2004	25033	10026	40	5	3	3	60
2005	25357	10335	41	7	3	3	43
2006	N/A	N/A	N/A	1	1	0	0
2007	N/A	N/A	N/A	2	2	1	50
Total				80	1	30	38

B. Other

Year	Ph.D.s	Intending to be Postdoc	%	Applications	% of All	Accepted	%
1965	8419	2350	28	37	28	29	78
1966	9383	2714	29	47	25	30	64
1967	10415	3078	30	58	29	37	64
1968	11821	3635	31	71	20	41	58
1969	13366	4169	31	128	21	26	20
1970	15033	4831	32	105	13	26	25
1971	16919	5489	32	100	9	27	27
1972	18449	6058	33	119	9	21	18
1973	19560	6369	33	108	9	23	21
1974	20131	6495	32	207	17	27	13
1975	20315	6654	33	163	14	25	15
1976	20196	6824	34	142	14	24	17
1977	20124	7068	35	155	20	22	14
1978	20018	7445	37	203	25	29	14
1979	20272	8045	40	172	22	24	14
1980	20702	8475	41	136	19	30	22
1981	21086	8799	42	183	24	49	27
1982	21670	9079	42	149	19	39	26
1983	22022	9238	42	139	19	42	30
1984	22410	9369	42	161	19	55	34
1985	22543	9404	42	219	26	65	30
1986	22659	9537	42	254	29	90	35
1987	22719	9696	43	210	23	75	36
1988	23332	10068	43	237	27	108	46
1989	23820	10335	43	225	26	84	37
1990	24582	10813	44	257	30	89	35
1991	25696	11331	44	229	21	78	34
1992	26968	11915	44	220	21	79	36
1993	28264	12455	44	254	20	78	31
1994	29672	13044	44	248	21	98	40
1995	31093	13444	43	201	23	66	33
1996	32451	13876	43	149	20	76	51
1997	33749	14102	42	151	20	69	46
1998	34805	14243	41	132	23	63	48
1999	35295	14243	40	100	19	42	42
2000	36034	14349	40	99	20	55	56
2001	36220	14210	39	94	17	43	46
2002	36354	14238	39	111	15	58	52
2003	36335	14316	39	125	15	60	48
2004	37005	14737	40	110	15	46	42
2005	37706	15222	40	122	16	39	32
2006	N/A	N/A	N/A	85	20	39	46
2007	N/A	N/A	N/A	27	18	12	44
Total				6442	19	2138	33

Source: Doctoral pool data from NSF, Survey of Earned Doctorates; other data from NRC DataRAP database.

APPENDIX TABLE E-4 Doctoral Pool of Potential Postdocs, Applications, and Acceptances in Engineering, by Research Associateship Program

A. NIST

Year	Ph.D.s	Intending to be Postdoc	%	Applications	% of All	Accepted	%
1965	5564	276	5	7	11	6	86
1966	6529	341	5	6	9	3	50
1967	7527	382	5	10	11	9	90
1968	8592	392	5	7	6	2	29
1969	9706	411	4	13	7	1	8
1970	10647	415	4	23	11	3	13
1971	11383	452	4	39	15	3	8
1972	11789	501	4	52	22	5	10
1973	11838	570	5	16	10	4	25
1974	11195	562	5	15	11	2	13
1975	10396	562	5	17	12	2	12
1976	9528	574	6	22	17	6	27
1977	8670	568	7	14	13	3	21
1978	7777	516	7	18	17	8	44
1979	7313	517	7	15	14	3	20
1980	6850	488	7	13	16	5	38
1981	6462	423	7	15	17	7	47
1982	6158	369	6	10	11	1	10
1983	6060	361	6	20	15	7	35
1984	6005	375	6	16	15	4	25
1985	6028	396	7	16	16	5	31
1986	6240	446	7	5	10	N/A	N/A
1987	6626	540	8	33	27	6	18
1988	7242	610	8	25	20	6	24
1989	7868	688	9	21	20	7	33
1990	8546	779	9	30	22	1	3
1991	9250	867	9	42	21	5	12
1992	9802	941	10	67	30	6	9
1993	10250	1073	10	92	39	14	15
1994	10599	1191	11	87	34	10	11
1995	11028	1304	12	60	27	10	17
1996	11537	1411	12	49	31	20	41
1997	12165	1513	12	59	35	15	25
1998	12505	1505	12	25	21	12	48
1999	12777	1548	12	40	30	12	30
2000	12614	1547	12	33	28	16	48
2001	12171	1501	12	21	27	6	29
2002	11325	1434	13	35	24	17	49
2003	10668	1466	14	53	32	18	34
2004	10126	1497	15	55	33	18	33
2005	9902	1588	16	91	42	23	25
2006	N/A	N/A	N/A	69	38	21	30
2007	N/A	N/A	N/A	32	30	16	50
Total				1388	23	348	25

B. Other

Year	Ph.D.s	Intending to be Postdoc	%	Applications	% of All	Accepted	%
1965	7251	378	5	7	5	7	100
1966	8612	474	6	16	8	16	100
1967	10000	538	5	24	12	24	100
1968	11498	584	5	21	6	13	62
1969	13099	638	5	63	10	16	25
1970	14459	685	5	145	18	41	28
1971	15656	785	5	376	33	63	17
1972	16555	954	6	417	32	76	18
1973	17064	1108	6	305	26	60	20
1974	16946	1155	7	262	22	61	23
1975	16514	1194	7	297	25	60	20
1976	15850	1244	8	253	26	46	18
1977	14990	1192	8	144	19	27	19
1978	14049	1144	8	150	18	26	17
1979	13392	1154	9	144	18	36	25
1980	12869	1158	9	118	17	38	32
1981	12563	1081	9	138	18	43	31
1982	12566	1041	8	152	19	35	23
1983	12924	1032	8	155	21	37	24
1984	13346	1090	8	208	25	45	22
1985	14033	1155	8	150	18	30	20
1986	14880	1336	9	148	17	49	33
1987	15946	1560	10	181	20	42	23
1988	17351	1819	10	164	19	48	29
1989	18982	2074	11	173	20	58	34
1990	20710	2314	11	157	18	40	25
1991	22548	2518	11	247	23	68	28
1992	24274	2733	11	232	23	65	28
1993	25786	2953	11	340	26	81	24
1994	27063	3124	12	289	25	74	26
1995	28177	3338	12	204	24	63	31
1996	29273	3537	12	176	24	56	32
1997	29949	3628	12	159	21	53	33
1998	30172	3624	12	111	19	40	36
1999	29682	3627	12	126	24	39	31
2000	28997	3573	12	117	23	48	41
2001	28196	3445	12	136	25	82	60
2002	27159	3486	13	214	29	128	60
2003	26517	3741	14	278	34	91	33
2004	26962	4187	16	202	28	43	21
2005	28043	4682	17	201	27	40	20
2006	N/A	N/A	N/A	140	33	43	31
2007	N/A	N/A	N/A	43	29	10	23
Total				7583	23	2061	27

Source: Doctoral pool data from NSF, Survey of Earned Doctorates; other data from NRC DataRAP database.

APPENDIX TABLE E-5 Doctoral Pool of Potential Postdocs, Applications, and Acceptances in Mathematical and Computer Sciences, by Research Associateship Program

A. NIST

Year	Ph.D.s	Intending to be Postdoc	%	Applications	% of All	Accepted	%
1965	2040	164	8	4	6	4	100
1966	2400	183	8	3	4	3	100
1967	2776	183	7	1	1	1	100
1968	3175	181	6	6	6	1	17
1969	3576	188	5	11	6	1	9
1970	4031	203	5	6	3	2	33
1971	4413	205	5	13	5	2	15
1972	4764	213	4	16	7	1	6
1973	4926	227	5	11	7	2	18
1974	4914	214	4	12	9	3	25
1975	4733	201	4	18	13	4	22
1976	4469	195	4	10	8	0	0
1977	4153	188	5	8	7	1	13
1978	3898	182	5	5	5	2	40
1979	3737	194	5	6	6	2	33
1980	3565	207	6	1	1	1	100
1981	3467	220	6	5	6	0	0
1982	3348	243	7	2	2	0	0
1983	3238	253	8	0	0	N/A	N/A
1984	3109	263	8	2	2	0	0
1985	2998	259	9	3	3	1	33
1986	2916	276	9	3	6	N/A	N/A
1987	2904	278	10	1	1	0	0
1988	2936	287	10	1	1	0	0
1989	3082	311	10	2	2	0	0
1990	3241	347	11	2	1	1	50
1991	3522	384	11	8	4	2	25
1992	3811	427	11	8	4	2	25
1993	4108	479	12	8	3	1	13
1994	4306	513	12	14	6	6	43
1995	4619	560	12	13	6	4	31
1996	4681	588	13	8	5	2	25
1997	4763	598	13	6	4	2	33
1998	4900	612	12	6	5	1	17
1999	4930	629	13	4	3	2	50
2000	4799	626	13	4	3	2	50
2001	4727	641	14	0	0	N/A	N/A
2002	4535	666	15	5	3	2	40
2003	4335	717	17	5	3	1	20
2004	4228	791	19	3	2	0	0
2005	4206	873	21	4	2	1	25
2006	N/A	N/A	N/A	4	2	2	50
2007	N/A	N/A	N/A	6	6	2	33
Total				258	4	64	25

B. Other

Year	Ph.D.s	Intending to be Postdoc	%	Applications	% of All	Accepted	%
1965	2476	198	8	3	2	3	100
1966	2913	222	8	5	3	4	80
1967	3355	231	7	5	2	5	100
1968	3843	234	6	33	9	20	61
1969	4325	260	6	17	3	3	18
1970	4865	284	6	25	3	7	28
1971	5334	296	6	34	3	4	12
1972	5785	319	6	47	4	5	11
1973	6047	348	6	58	5	11	19
1974	6188	331	5	51	4	6	12
1975	6110	330	5	65	6	8	12
1976	5875	318	5	35	4	7	20
1977	5558	302	5	18	2	3	17
1978	5284	299	6	26	3	10	38
1979	5052	311	6	16	2	1	6
1980	4867	321	7	17	2	2	12
1981	4824	350	7	29	4	5	17
1982	4800	380	8	24	3	8	33
1983	4828	408	8	23	3	4	17
1984	4842	437	9	17	2	3	18
1985	4878	455	9	26	3	5	19
1986	5046	520	10	32	4	9	28
1987	5295	572	11	26	3	6	23
1988	5572	611	11	22	3	6	27
1989	6050	682	11	15	2	5	33
1990	6649	785	12	38	4	16	42
1991	7359	881	12	40	4	9	23
1992	8097	995	12	36	4	8	22
1993	8859	1101	12	66	5	13	20
1994	9409	1177	13	38	3	10	26
1995	9999	1269	13	25	3	7	28
1996	10203	1332	13	34	5	10	29
1997	10308	1348	13	34	5	10	29
1998	10386	1390	13	22	4	3	14
1999	10304	1452	14	27	5	9	33
2000	10027	1459	15	18	4	4	22
2001	9817	1488	15	28	5	15	54
2002	9511	1596	17	35	5	19	54
2003	9266	1737	19	42	5	11	26
2004	9351	1903	20	31	4	9	29
2005	9780	2128	22	20	3	7	35
2006	N/A	N/A	N/A	27	6	9	33
2007	N/A	N/A	N/A	11	7	4	36
Total				1241	4	323	26

Source: Doctoral pool data from NSF, Survey of Earned Doctorates; other data from NRC DataRAP database.

APPENDIX TABLE E-6 Doctoral Pool of Potential Postdocs, Applications, and Acceptances in the Physical Sciences, by Research Associateship Program.

A. NIST

Year	Ph.D.s	Intending to be Postdoc	%	Applications	% of All	Accepted	%
1965	10049	2089	21	53	83	51	96
1966	10867	2414	22	60	86	30	50
1967	11932	2736	23	82	88	43	52
1968	12964	2984	23	95	88	12	13
1969	14098	3364	24	165	87	13	8
1970	15338	3838	25	185	86	10	5
1971	16536	4354	26	198	79	11	6
1972	17050	4772	28	170	71	13	8
1973	17147	5077	30	126	82	14	11
1974	16696	5038	30	112	80	12	11
1975	15874	4911	31	109	76	11	10
1976	14854	4685	32	97	75	17	18
1977	14110	4402	31	87	78	18	21
1978	13442	4247	32	80	77	13	16
1979	13213	4262	32	82	78	17	21
1980	12791	4135	32	67	83	19	28
1981	12533	3950	32	66	76	15	23
1982	12422	3858	31	74	82	15	20
1983	12475	3821	31	109	83	17	16
1984	12452	3862	31	86	80	21	24
1985	12543	3932	31	81	79	22	27
1986	12552	4055	32	40	83	N/A	N/A
1987	12534	4297	34	87	71	16	18
1988	12593	4541	36	102	80	16	16
1989	12546	4653	37	82	78	13	16
1990	12744	4792	38	100	75	21	21
1991	13020	4966	38	148	74	19	13
1992	13177	5032	38	148	65	22	15
1993	13122	5006	38	137	58	19	14
1994	13324	5079	38	152	60	24	16
1995	13254	5121	39	144	65	24	17
1996	13080	5033	38	95	61	33	35
1997	13085	4948	38	103	61	23	22
1998	13179	4906	37	84	70	42	50
1999	12984	4754	37	84	64	23	27
2000	12735	4642	36	75	64	34	45
2001	12494	4554	36	58	73	29	50
2002	12055	4462	37	99	69	31	31
2003	11698	4392	38	102	62	35	34
2004	11343	4351	38	103	62	27	26
2005	11166	4356	39	116	53	32	28
2006	N/A	N/A	N/A	109	60	36	33
2007	N/A	N/A	N/A	67	63	28	42
Total				4419	72	941	21

B. Other

Year	Ph.D.s	Intending to be Postdoc	%	Applications	% of All	Accepted	%
1965	11909	2533	21	82	62	72	88
1966	12975	2910	22	121	63	85	70
1967	14381	3349	23	111	55	80	72
1968	15636	3674	23	221	63	103	47
1969	17044	4162	24	389	65	91	23
1970	18582	4766	26	515	65	97	19
1971	20024	5401	27	607	54	120	20
1972	20778	5920	28	727	55	150	21
1973	21174	6358	30	703	59	108	15
1974	21004	6416	31	685	56	111	16
1975	20311	6332	31	631	54	133	21
1976	19316	6136	32	550	56	114	21
1977	18474	5809	31	449	58	88	20
1978	17630	5590	32	443	54	111	25
1979	17185	5580	32	444	57	119	27
1980	16624	5414	33	423	61	124	29
1981	16328	5188	32	388	52	120	31
1982	16264	5095	31	457	58	113	25
1983	16468	5092	31	400	55	111	28
1984	16607	5164	31	432	52	116	27
1985	16991	5333	31	421	51	137	33
1986	17460	5659	32	425	49	134	32
1987	17949	6197	35	482	53	136	28
1988	18555	6732	36	427	50	144	34
1989	19080	7151	37	455	52	164	36
1990	19809	7544	38	393	46	139	35
1991	20571	7945	39	557	52	160	29
1992	21305	8217	39	528	52	137	26
1993	21731	8286	38	617	48	178	29
1994	22548	8524	38	595	50	184	31
1995	22907	8671	38	419	49	133	32
1996	23099	8659	37	365	49	146	40
1997	23171	8495	37	376	51	142	38
1998	23338	8500	36	310	53	113	36
1999	22921	8246	36	262	50	114	44
2000	22467	8031	36	255	51	109	43
2001	21977	7840	36	282	52	150	53
2002	21318	7772	36	359	49	187	52
2003	20789	7762	37	356	44	142	40
2004	20429	7881	39	355	50	117	33
2005	20622	8240	40	402	53	81	20
2006	N/A	N/A	N/A	171	40	53	31
2007	N/A	N/A	N/A	65	43	24	37
Total				17655	53	5190	29

Source: Doctoral pool data from NSF, Survey of Earned Doctorates; other data from NRC DataRAP database.

APPENDIX TABLE E-7 Doctoral Pool of Potential Postdocs, Applications, and Acceptances, by Gender and Research Associateship Program

A. NIST

	Female Doctoral Pool (%)		Applications			Accepted		
Year	Ph.D.s	Intending to be Postdocs	Women	Men	Women (%)	Women	Men	Women (%)
1965	5	8	2	62	3	2	59	3
1966	5	8	1	69	1	0	36	0
1967	6	9	2	91	2	0	53	0
1968	6	9	1	107	1	1	14	7
1969	6	10	7	183	4	1	15	6
1970	7	10	5	209	2	1	14	7
1971	7	10	3	249	1	0	16	0
1972	7	11	6	232	3	0	19	0
1973	8	11	8	145	5	2	18	10
1974	9	12	7	133	5	0	17	0
1975	10	13	12	132	8	1	16	6
1976	11	14	3	126	2	1	22	4
1977	12	16	5	106	5	1	21	5
1978	13	17	5	99	5	0	23	0
1979	14	18	10	95	10	2	22	8
1980	16	19	7	74	9	2	23	8
1981	17	21	7	80	8	1	21	5
1982	19	22	9	81	10	0	16	0
1983	20	24	12	119	9	1	23	4
1984	21	25	12	96	11	3	23	12
1985	23	26	18	84	18	6	22	21
1986	24	27	11	37	23	N/A	N/A	N/A
1987	25	27	12	110	10	3	19	14
1988	25	28	17	111	13	3	19	14
1989	27	29	16	89	15	5	15	25
1990	27	30	15	119	11	3	22	12
1991	28	31	21	179	11	3	23	12
1992	29	31	27	199	12	6	25	19
1993	30	32	28	209	12	4	30	12
1994	30	33	36	218	14	13	27	33
1995	31	33	34	188	15	7	31	18
1996	32	34	28	128	18	7	50	12
1997	32	35	34	136	20	10	30	25
1998	33	36	21	99	18	10	48	17
1999	33	36	23	109	17	5	34	13
2000	34	36	18	100	15	10	47	18
2001	35	37	20	59	25	13	22	37
2002	36	37	29	114	20	11	41	21
2003	37	38	30	135	18	11	45	20
2004	38	39	45	121	27	13	35	27
2005	39	39	40	178	18	16	43	27
2006	N/A	N/A	44	139	24	16	43	27
2007	N/A	N/A	30	77	28	13	34	28
Total			721	5426	12	207	1176	15

B. Other

Year	Female Doctoral Pool (%) Ph.D.s	Intending to be Postdocs	Applications Women	Men	Women (%)	Accepted Women	Men	Women (%)
1965	5	8	4	128	3	4	109	4
1966	6	9	5	186	3	2	133	1
1967	6	9	9	193	4	6	141	4
1968	6	9	5	345	1	3	176	2
1969	6	10	20	583	3	3	134	2
1970	7	10	16	779	2	2	170	1
1971	7	10	36	1095	3	11	205	5
1972	7	10	47	1270	4	8	245	3
1973	8	11	47	1140	4	5	197	2
1974	9	11	61	1153	5	9	196	4
1975	9	13	38	1127	3	9	219	4
1976	10	14	64	924	6	17	175	9
1977	11	15	70	706	9	14	126	10
1978	12	16	79	748	10	19	157	11
1979	13	17	84	695	11	15	165	8
1980	14	18	67	632	10	20	175	10
1981	15	19	87	666	12	28	193	13
1982	16	21	77	711	10	18	177	9
1983	17	22	86	646	12	32	164	16
1984	18	23	102	724	12	30	190	14
1985	19	24	120	707	15	43	196	18
1986	20	24	112	755	13	42	242	15
1987	21	25	136	771	15	54	206	21
1988	21	25	144	718	17	65	245	21
1989	22	26	139	742	16	60	255	19
1990	22	26	132	717	16	47	239	16
1991	23	27	192	887	18	61	257	19
1992	23	27	157	867	15	54	237	19
1993	24	28	219	1069	17	74	281	21
1994	24	29	176	1003	15	70	299	19
1995	25	29	150	713	17	59	216	21
1996	26	30	122	617	17	60	233	20
1997	27	31	132	611	18	60	223	21
1998	28	31	124	456	21	59	161	27
1999	28	32	105	418	20	46	161	22
2000	29	32	105	397	21	51	171	23
2001	30	33	117	430	21	64	229	22
2002	31	34	162	568	22	91	307	23
2003	32	34	167	641	21	69	239	22
2004	33	35	184	528	26	63	153	29
2005	33	35	186	569	25	50	120	29
2006	N/A	N/A	109	319	25	47	99	32
2007	N/A	N/A	40	110	27	13	37	26
Total			4234	29064	13	1557	8253	16

Source: Doctoral pool data from NSF, Survey of Earned Doctorates; other data from NRC DataRAP database.

APPENDIX TABLE E-8 Doctoral Pool of Potential Postdocs, Applications, and Acceptance by Race/Ethnicity and Research Associateship Program

A. NIST

	Doctoral Pool		Applications					Accepted				
Year	Ph.D. Minority (%)	URM Intending to be Postdoc (%)	White	URM	Asian	Unk.	URM (%)	White	URM	Asian	Unk.	URM (%)
1980	3	1	66	3	5	7	4	23	1	0	1	4
1981	3	2	74	2	1	10	3	21	0	0	1	0
1982	3	2	5	0	0	85	0	0	0	0	16	N/A
1983	3	2	103	2	2	24	2	18	0	1	5	0
1984	3	2	86	0	2	20	0	23	0	0	3	0
1985	3	2	75	2	3	22	3	24	0	1	3	0
1986	3	2	37	1	2	8	3	N/A	N/A	N/A	N/A	N/A
1987	3	3	84	5	4	29	5	15	0	1	6	0
1988	3	3	105	1	5	17	1	19	0	1	2	0
1989	4	3	98	0	3	4	0	20	0	0	0	0
1990	4	3	114	1	8	11	1	22	0	1	2	0
1991	4	3	159	7	12	22	4	24	1	0	1	4
1992	4	4	189	1	14	22	0	30	0	1	0	0
1993	4	4	168	8	14	47	4	27	2	0	5	7
1994	5	4	166	6	12	70	3	27	1	1	11	3
1995	5	4	158	9	17	38	5	24	3	4	7	10
1996	5	5	96	3	8	49	3	38	2	3	14	5
1997	6	5	135	12	8	15	8	33	3	2	2	8
1998	6	5	85	9	9	17	9	39	4	6	9	8
1999	6	5	91	9	6	26	8	28	1	3	7	3
2000	7	6	86	11	9	12	10	42	6	4	5	12
2001	7	6	60	5	10	4	7	27	4	2	2	12
2002	7	7	120	7	10	6	5	43	4	4	1	8
2003	8	7	142	5	12	6	3	50	1	3	2	2
2004	8	7	140	10	13	3	6	43	3	2	0	6
2005	8	8	180	6	28	4	3	49	1	8	1	2
2006	N/A	N/A	154	10	14	5	6	49	3	5	2	5
2007	N/A	N/A	87	6	11	3	6	39	3	5	0	6
Total			3063	141	242	586	4	797	43	58	108	5

B. Other

Year	Doctoral Pool: Ph.D. Minority (%)	Doctoral Pool: URM Intending to be Postdoc (%)	Applications: White	Applications: URM	Applications: Asian	Applications: Unk.	Applications: URM (%)	Accepted: White	Accepted: URM	Accepted: Asian	Accepted: Unk.	Accepted: URM (%)
1980	5	2	284	7	11	397	2	97	2	3	93	2
1981	6	3	356	4	8	385	1	125	2	3	91	2
1982	6	3	57	1	0	730	2	20	0	0	175	0
1983	6	3	245	4	11	472	2	86	2	5	103	2
1984	6	3	311	5	21	489	1	117	0	6	97	0
1985	6	3	283	4	20	520	1	103	2	5	129	2
1986	6	4	346	8	16	497	2	129	2	6	147	1
1987	6	4	308	9	9	581	3	111	5	4	140	4
1988	6	4	403	1	19	439	0	175	1	9	125	1
1989	6	5	380	7	12	482	2	162	3	6	144	2
1990	6	5	400	19	64	366	4	154	3	9	120	2
1991	6	5	502	25	88	464	4	167	7	19	125	4
1992	6	5	465	25	81	453	4	163	6	23	99	3
1993	6	5	456	21	45	766	4	177	4	13	161	2
1994	6	5	378	13	70	718	3	157	4	21	187	2
1995	6	5	301	23	61	478	6	119	7	17	132	5
1996	6	5	291	31	72	345	8	144	6	33	110	3
1997	6	5	468	31	118	126	5	193	8	45	37	3
1998	7	5	316	22	79	163	5	139	10	25	46	6
1999	7	5	247	18	79	179	5	110	3	29	65	2
2000	7	6	313	21	66	102	5	144	10	28	40	5
2001	7	6	370	29	100	48	6	206	16	49	22	6
2002	8	7	449	42	163	76	6	264	22	79	33	6
2003	8	7	464	34	253	57	5	212	11	62	23	4
2004	8	7	434	28	228	22	4	163	6	42	5	3
2005	8	8	414	30	270	41	4	121	6	37	6	4
2006	N/A	N/A	237	23	153	15	6	106	2	35	3	1
2007	N/A	N/A	99	4	43	4	3	43	0	5	2	0
Total			9577	489	2160	9415	4	3907	150	618	2460	3

Source: Doctoral pool data from NSF, Survey of Earned Doctorates; other data from NRC DataRAP database.

APPENDIX TABLE E-9 Applicant's Marital Status, by Research Associateship Program

	NIST/NRC RAP			OTHER RAP			NIST (%)	Other (%)
Year	Married	Single	Total	Married	Single	Total	Married	Married
1965	40	24	64	97	35	132	63	73
1966	59	11	70	135	53	188	84	72
1967	68	24	92	154	41	195	74	79
1968	72	32	104	242	83	325	69	74
1969	137	49	186	419	143	562	74	75
1970	168	46	214	605	178	783	79	77
1971	191	61	252	846	278	1124	76	75
1972	173	65	238	964	353	1317	73	73
1973	113	40	153	879	308	1187	74	74
1974	102	38	140	889	325	1214	73	73
1975	98	46	144	875	290	1165	68	75
1976	82	47	129	687	301	988	64	70
1977	71	40	111	537	239	776	64	69
1978	59	45	104	524	303	827	57	63
1979	50	55	105	509	268	777	48	66
1980	40	41	81	460	239	699	49	66
1981	41	46	87	488	265	753	47	65
1982	38	51	89	506	277	783	43	65
1983	57	74	131	494	234	728	44	68
1984	46	62	108	540	283	823	43	66
1985	42	60	102	504	321	825	41	61
1986	21	27	48	547	315	862	44	63
1987	60	62	122	579	317	896	49	65
1988	56	71	127	556	297	853	44	65
1989	45	60	105	544	334	878	43	62
1990	62	72	134	563	282	845	46	67
1991	94	106	200	685	389	1074	47	64
1992	104	122	226	669	353	1022	46	65
1993	114	120	234	859	427	1286	49	67
1994	117	137	254	751	426	1177	46	64
1995	107	114	221	503	360	863	48	58
1996	76	79	155	468	269	737	49	64
1997	70	99	169	456	286	742	41	61
1998	44	75	119	333	241	574	37	58
1999	60	71	131	314	204	518	46	61
2000	58	60	118	300	199	499	49	60
2001	29	50	79	332	215	547	37	61
2002	53	90	143	479	249	728	37	66
2003	66	98	164	513	288	801	40	64
2004	83	83	166	395	314	709	50	56
2005	104	113	217	431	318	749	48	58
2006	77	106	183	261	165	426	42	61
2007	41	66	107	77	73	150	38	51
Total	3288	2838	6126	21969	11138	33107	54	66

APPENDIX TABLE E-10 Awardees' Marital Status, by Research Associateship Program

	NIST/NRC RAP			OTHER RAP			NIST (%)	Other (%)
Year	Married	Single	Total	Married	Single	Total	Married	Married
1965	37	24	61	80	33	113	61	71
1966	31	5	36	95	37	132	86	72
1967	41	12	53	109	32	141	77	77
1968	9	6	15	123	37	160	60	77
1969	11	2	13	98	35	133	85	74
1970	8	7	15	136	34	170	53	80
1971	13	3	16	164	51	215	81	76
1972	15	4	19	194	59	253	79	77
1973	14	6	20	153	49	202	70	76
1974	13	4	17	164	41	205	76	80
1975	12	5	17	169	59	228	71	74
1976	15	8	23	129	63	192	65	67
1977	13	9	22	101	39	140	59	72
1978	12	11	23	122	54	176	52	69
1979	14	10	24	127	51	178	58	71
1980	13	12	25	133	62	195	52	68
1981	9	13	22	142	79	221	41	64
1982	5	10	15	123	72	195	33	63
1983	9	15	24	120	75	195	38	62
1984	10	16	26	137	82	219	38	63
1985	14	14	28	138	99	237	50	58
1986	N/A	N/A	N/A	174	108	282	N/A	62
1987	11	11	22	156	104	260	50	60
1988	7	15	22	198	112	310	32	64
1989	11	9	20	196	119	315	55	62
1990	10	15	25	195	91	286	40	68
1991	16	10	26	191	126	317	62	60
1992	13	18	31	177	112	289	42	61
1993	16	18	34	215	140	355	47	61
1994	16	24	40	229	140	369	40	62
1995	15	23	38	154	121	275	39	56
1996	27	30	57	191	102	293	47	65
1997	15	25	40	162	121	283	38	57
1998	20	38	58	135	84	219	34	62
1999	16	23	39	129	78	207	41	62
2000	28	29	57	133	87	220	49	60
2001	10	25	35	187	106	293	29	64
2002	19	33	52	271	127	398	37	68
2003	23	32	55	197	110	307	42	64
2004	26	22	48	125	90	215	54	58
2005	26	33	59	96	74	170	44	56
2006	27	32	59	94	52	146	46	64
2007	19	28	47	21	29	50	40	42
Total	689	689	1378	6383	3376	9759	50	65

Appendix F
PH.D. FIELDS BY BROAD CATEGORY

Agriculture sciences and natural resources is: Agribusiness; Agricultural Entomology; Agricultural Management; Agricultural Science; Agricultural Technology; Agriculture; Agronomy; Agronomy and Genetics; Agronomy and Soil Sciences; Animal Agriculture; Animal Breeding; Animal Diseases; Animal Husbandry; Animal Industry; Animal Nutrition; Aquaculture; Avian Science; Coastal Zone Management; Conservation; Crop and Soil Science; Dairy Husbandry; Dairy Industry; Dairy Science; Entomology Parasitology; Farm & Ranch Management; Farm Crops; Farm Management; Feed Technology; Field Crops; Fire Science; Fish and Game; Fish and Wildlife; Fish, Game, and Wildlife Management; Fisheries; Floriculture; Flour and Feed Mill Industry; Forest Botany; Forest Chemistry; Forest Ecology; Forest Economics; Forest Entomology; Forest Management; Forest Products; Forest Resources; Forest Zoology; Forestry; Forestry and Horticulture; Game Management; General Agriculture; Horticulture; Irrigation; Land Use Management & Reclaimation; Mechanized Agriculture; Natural Resource Management; Natural Resources; Paper & Pulp Science; Park Management; Pest Control Technology; Plant Agriculture; Plant Breeding; Plant Industry; Pomology; Poultry; Poultry Husbandry; Poultry Industry; Poultry Science; Pulp and Paper; Range Ecology; Range Management; Range Science; Recreation and Park Management; Recreation and Parks; Resource Development; Resource Sciences; Silviculture; Soil and Water Science; Soil Chemistry; Soil Microbiology; Soil Physics; Soil Science; Soils; Soils and Meteorology; Subtropical Horticulture; Vegetable Crops; Water Resources; Watershed Management; Wildlife Wildlife Conservation; Wildlife Management; Wildlife Technology; Wood Chemistry; Wood Science; and Wood Technology.

Biological, biomedical, and health sciences includes: Acupuncture and Oriental Medicine; Adult Nursing; Advanced Practice Nursing; Agricultural Biochemistry; Agricultural Biochemistry; Nutrition; Agricultural Microbiology; Aids; Allergy; Allied Health; Allopathic Medicine; Anatomy; Anatomy & Cell Biology; Anesthesiology; Animal Behavior; Animal Biology; Animal Ecology; Animal Genetics; Animal Pathology; Animal Physiology; Animal Science; Applied Biology; Arthritis; Audiology; Aviation Medicine; Bacteriology; Bacteriology and Medical Technology; Bacteriology Public Health; Behavioral Biology; Behavioral Genetics; Biobehavioral Sciences; Biochemical Pharmacology; Biochemical Science; Biochemical Technology; Biochemistry; Biochemistry and Nutrition; Biochemistry Biophysics; Bioethics; Biological and Biomedical Science; Biological Chemistry; Biological Structure; Biology; Biology and Genetics; Biomedical Science; Biomedical Technologies; Bionucleonics; Biophysical Chemistry; Biophysical Sciences; Biophysics; Biophysics Microbiology; Biophysics Physics Biochemistry; Botanical Science; Botany; Botany and Microbiology; Botany and Plant Pathology; Botany and Plant Sciences; Cardiology; Cardiovascular Medicine; Cardiovascular Sciences; Cardiovascular Surgery; Cell Physiology; Cellular Biology; Chemical Biology; Chest Diseases; Clinical Microbiology; Comp Biochem Physiology; Comparative Biochemistry; Comparative Physiology; Cytology; Dental Hygiene; Dentistry; Dermatology; Dermatology Syphilology; Developmental Biology; Ecology; Economic Zoology; Embryology; Emergency Medicine; Endocrinology; Entomology; Environmental Biology; Environmental Health; Environmental Medicine; Environmental Stress; Epidemiology; Eugenics; Evolutionary Biology;

Exobiology; Experimental Biology; Experimental Medicine; Experimental Pathology; Family Practice; Food and Nutrition; Food Science; Food Science Technology; Food Technology; Foods; Fungus Physiology; Gastroenterology; General Practice; Genetics; Gerontology; Gynecology; Hematology; Histology; History Of Medicine; Human Anatomy; Human Biology; Human Ecology; Human Genetics; Human Reproduction; Hydrobiology; Hygiene; Immunology; Immunoparasitology; Infections; Infectious Diseases; Insect Biology; Internal Medicine; Intl Agricultural Dev; L/Sci Othr; Legal Medicine; Life Science; Limnology; Marine Biology; Marine Microbiology; Marine Science; Medical Biochemistry; Medical Micro Immunology; Medical Microbiology; Medical Physics; Medical Physiology; Medical Research; Medical Technology; Medicinal Chemistry; Medicine; Metabolism; Microbial Genetics; Microbiology; Microscopic Anatomy; Molecular Basis Biol Phenom; Molecular Biology; Molecular Immunology; Molecular Medicine; Molecular Neurobiology; Molecular Pathology; Molecular Virology; Morphology; Natural Sciences; Naturopathic Medicine; Neural Prosthetics; Neuroanatomy; Neurobiology; Neurochemistry; Neurocommunications; Neuroendocrinology; Neurological Surgery; Neurology; Neuropharmacology; Neurophysiology; Neurosciences; Neurosurgery; Neurotoxicology; Nmr Imaging; Nuclear Medical Technology; Nuclear Medicine; Nurse Anesthesia; Nurse Midwifery; Nursing; Nutrition; Nutrition and Metabolism; Obstetrics; Obstetrics and Gynecology; Occupational Health; Oncology; Ophthalmology; Oral Pathology; Oral Surgery; Organismic Biology; Ornithology; Orthodontics; Orthopedic Surgery; Orthopedics; Otolaryngology; Paleozoology; Parasitology; Pathobiology; Pathology; Pathophysiology; Pediatric Nursing; Pediatrics; Pedodontics; Periodontology; Pharmaceutics; Pharmacognosy; Pharmacology; Pharmacology Toxicology; Pharmacy; Physical Medicine; Physical Medicine and Rehab; Physiological Chem; Physiological Hygiene; Physiological Science; Physiology; Physiology and Anatomy; Physiology and Biophysics; Physiology Pharmacology; Phytochemistry; Phytopathology; Plant and Soil Science; Plant Biochemistry; Plant Biology; Plant Genetics; Plant Molecular Biology; Plant Nematology; Plant Nutrition; Plant Path Bacteriology; Plant Pathology; Plant Physiology; Plant Science; Plastic Surgery; Podiatric Medicine; Population Biology; Population Environ Biol; Postgraduate Medicine; Practical Nursing; Prev Med Public Health; Preventive Medicine; Proctology; Psychiatric Nursing; Psychiatry; Psychiatry and Neurology; Psychobiology; Public Health; Radiation Biochemistry; Radiation Biol and Biophys; Radiation Biology; Radiation Biophysics; Radiation Medicine; Radiation Physiology; Radiobiology; Radiological Physics; Radiology; Radiopharmacy; Reproductive Biology; Reproductive Physiology; Respiratory Therapy; Sheep Science; Speech Pathology; Structural Biology; Surgery; Surgical Nursing; Systematic Biology; Theoretical Biology; Thoracic Surgery; Toxicokinetics; Toxicology; Traumatology; Tropical Medicine; Urology; Vertebrate Zoology; Vet Public Health; Veterinary Anatomy; Veterinary Bacteriology; Veterinary Biochem; Veterinary College; Veterinary Histology; Veterinary Medicine; Veterinary Parasitology; Veterinary Pathology; Veterinary Pharmacology; Veterinary Physiology; Veterinary Science; Viral Immunology; Virology; Virology and Epidemiology; Vision Sciences; Wildlife Biology; Zoology; Zoology & Oceanography; Zoology and Entomology; and Zoology and Physiology.

Engineering includes: Aero Engr & Engr Physics; Aero Safety Management; Aero/Astro Engineering; Aerodynamics; Aeronautical Admininistration; Aeronautical Engineering; Aeronautics; Aeronautics & Engr Mech; Aeronautics/Astronautics; Aeronomy; Aerophysics; Aerospace; Aerospace Engineering; Aerospace Mech Engr; Aerospace Science;

Aerospace/Aeronautical Engineering; Agricultural Engineering; Agricultural Irrigation Eng; Air Cond and Refrig Engr; Applied Analysis; Applied Mechanics; Architectural Engr; Architectural Technology; Astronautical Engineering; Astronautics; Automotive Engineering; Automotive Technologies; Aviation; Bio/Engr; Biochemical Engineering; Bioengineering; Biological Engineering; Biomedical Engineering; Biomedical Engr & Math; Biomolecular Engineering; Building Research; Ceramic Engineering; Ceramic Sciences and Engineering; Ceramic Technology; Ceramics; Chemical and Paper Engr; Chemical Engineering; Chemical Engr and Mat Sci; Chemistry and Metallurg Engr; Chemistry and Nuclear Engr; City Planning; Civil and Environ Engr; Civil and Geological Engr; Civil Engineering; Civil Engr and Engr Mech; Civil Engr Hydraulics; Communications; Computer Engineering; Construction Engineering and M; Economics Of Engineering; Electrical Computer Sci; Electrical Engineering; Electrical Engineering Technol; Electricity; Electromechanical Engineering; Electroncs; Electronic Engineering; Electronic Materials; Electronics and Instrument; Electro-Optics; Energy and Power Engineering; Energy Engineering; Eng/Indus; Eng/Metall; Eng/Mining; Engineering; Engineering Acoustics; Engineering Admin; Engineering Analysis; Engineering and Applied Phys; Engineering and Applied Science; Engineering Coastal; Engineering Design; Engineering Drawing; Engineering Graphics; Engineering Management; Engineering Mathematics; Engineering Mechanics; Engineering Phys and Mat Sci; Engineering Phys and Physics; Engineering Physics; Engineering Production; Engineering Science; Engineering Technology; Environmental and Sanitry Engr; Environmental Design; Environmental Engr; Environmental Sciences Engr; Fire Protection Engineering; Fluid & Thermal Sciences; Fluid Dynam and App Math; Fluid Dynamics; Fracture Mechanics; Fuel Technology; Fuels Engineering; Fundamental Sciences; General Engineering; Geological Engineering; Geology and Geological Engr; Geophys/Geophysical Engr; Geophysical Engr; Geophysics Engineering; Geotechnical Engineering; Glass Technology; Highway Engineering; Hydraulic Engineering; Hydraulics Industrial Apps Radiation; Industrial Communctn Engr; Industrial Engineering; Industrial Engr Mgmt Sci; Industrial Engr Operation Res; Industrial Management; Industrial Technology; Industrial/Management Engineer; Information Engr; Instrumentation Engr; Landscape Architecture; Machine Design; Management Engineering; Manufacturing Engr; Marine Engineering; Marine Engr Naval Arch; Material Science; Materials; Materials Engineering; Materials Sciences; Mech Aeroengr Mat Sci; Mech and Industrial Engr; Mech Engr Applied Mech; Mechanical Drawing; Mechanical Engineering; Mechanical Industries; Mechanical Science; Mechanics; Mechanics and Hydraulics; Metallurgical Engr; Metallurgical Technology; Metallurgy; Metallurigical Matrls Engr; Microelectronics; Mineral Dressing; Mineral Engineering; Mineral Industries; Mineral Preparation; Mineral Technology; Mineral/Mining Engineering; Mining; Mining and Metallurgy; Mining Engineering; Mining Geological Engr; Mining Technology; Natural Gas Engr; Nautical Science; Naval Architecture; Naval Sciences; Nuclear Engineering; Ocean Engineering; Optical Engineering; Organizational Sciences; Paper and Pulp Engineering; Petroleum and Chem Engr; Petroleum Engineering; Petroleum Production; Petroleum Refning Engr; Petroleum Science; Plastics; Plastics Engineering; Polymer Science and Engr; Radio Engineering; Reactor Technology; Reliability Engineering; Remote Sensing; Safety Engineering; Sanitary Engineering; Sanitary Engineering Technology; Sanitation; Sanitation Water Resource Engr; Science Engineering; Sensory Communications; Software Engineering; Solar Engineering; Solid State Sci and Tech; Space Engineering; Structural Engineering; Surveying; Surveying Science and Engineer; Systems Engineering;

Technical Sciences; Technology; Telecommunication Engineering; Textile Engineering; Textiles; Theoretical Applied Mech; Thermal Engineering; Traffic Engineering; Transportation; Transportation and Highway Eng; Transportation Engr; Water Engineering; Water Resources Engineering Welding Engineering; Wind Engineering; and Wood Products Engr.

Mathematical and computer sciences includes: Actuarial Science; Algebra; Analysis; Applied Math and Computer Science; Applied Mathematics; Applied Statistics; Artificial Intelligence; Biomathematics; Biometry; Biostatistics; Computation Machines; Computational Math; Computational Sciences; Computer Programming; Computer Science; Computer Technologies; Economic Statistics; Exact Science; Experimental Statistics; Functional Analysis; Geometry; Information Science; Information Studies; Logic; Math & Chemistry; Math and Applied Math; Math Biostatistics; Math/Appns; Mathematical Biology; Mathematical Sciences; Mathematical Statistics; Mathematics; Mathematics and Astronomy; Mathematics and Statistics; Operations; Operations Research; Prob&Stat; Quantitative Analysis; Quantitative Studies; Robotics; Statistics; Statistics and Computer Science; Technology Mathematics; Theoretical Statistics; and Topology.

Physical sciences includes: Acoustics; Aero & Planetary Atmos; Agricultural Chemistry; Air Pollution; Analytical Chemistry; Applied Chemistry; Applied Physics; Astrogeophysics; Astronomy; Astronomy Space Science; Astrophysics; Atmospheric and Space Sci; Atmospheric Chemistry; Atmospheric Physics; Atmospheric Sciences; Atom&Molec; Atomic Physics; Ballistics; Biogeochemistry; Biomedicinal Chemistry; Bio-Organic Chemistry; Biopharmaceutical Sci; Ceramic Chemistry; Chem Pharmaceutical Chem; Chem/Agr&F; Chem/Crmc; Chem/Poly; Chemical Physics; Chemical Technology; Chemistry; Chemistry and Physics; Climatology; Coatings Technology; Cosmic Rays; Cryogenics; Crystallography; Earth Planetary Science; Earth Sciences; Earth/Env; Economic Geology; Electromag; Electronics; Electrophysics; Elem Partl; Engineering Geology; Environmental Chemistry; Environmental Sciences; Experimental Physics; Flight Dynamics; Forecasting; Geochemistry; Geochronology; Geodetic Science; Geography; Geography and Anthropology; Geography and Meteorology; Geological Science; Geology; Geology & Geophysics; Geology and Geography; Geophys and Planetary Phys; Geophys Fluid Dynamics; Geophysical Institute; Geophysics; Geosciences; Gravitation; High Energy Physics; Holography; Hydrodynamics; Hydrography; Hydrology; Immunochemistry; Industrial Chemistry; Information Theory; Inorganic Chemistry; Institute Of Physics; Ionospheric Physics; Laser Physics; Lasers; Macromolecular Science; Magnetism; Marine Geology; Marine Technology; Maritime Sciences; Mathematical Physics; Meteorology; Meteorology Oceanography; Microbiochemistry; Mineral Economics; Mineral Science; Mineralogy; Molecular Biophy and Biochem; Molecular Biophysics; Molecular Physics; Nuclear Chemistry; Nuclear Physics; Nuclear Science and Engr; Nuclear Studies; Nuclear Technology; Oceanography; Optical Sciences; Optics; Organic Chemistry; Paleobotany; Paleontology; Paper Technology; Petrography; Petroleum Geology; Petrology; Pharmaceutical Chemistry; Photogrammetry; Phys/Atms; Phys/Fluid; Phys/Hi Pr; Phys/Hi Vc; Phys/Mech; Physical Chemistry; Physical Meteorology; Physical Oceanography; Physical Sciences; Physics; Physics and Astronomy; Physics and Astrophysics; Physics and Engr Physics; Physics and Geophysics; Physics and Mathematics; Planetary Science; Planetary Space Science; Plasma Physics; Polymer Chemistry; Polymer Science; Quantum Electrons; Quantum St; Rad Astron;

Radiochemistry; Radiophysics; Sanitary Chemistry; Sedimentary Structure; Seismology; Solar Physics; Solid State Physics; Solid State Science; Space Physics; Space Science; Spectroscopy; Stat Mech; Stratigraphy; Textile Chemistry; Theoretical Chemistry; Theoretical Physics; Theoretical Science; Therml Phn; Thermodyns; Tsunami; Water Chemistry; and Weather.

Appendix G

NAMES OF LABORATORIES

Lab Name on Application Form	New Name
Building and Fire Research Laboratory	Building and Fire Research Laboratory
Chemical Science and Technology Laboratory	Chemical Science and Technology Laboratory
Computer Systems Laboratory	Information Technology Laboratory
Computing and Applied Mathematics Laboratory	Information Technology Laboratory
Electronics & Electrical Engineering Laboratory	Electronics and Electrical Engineering Laboratory
Information Technology Laboratory	Information Technology Laboratory
Institute for Applied Technology	Multiple
Institute for Basic Standards	Multiple
Institute for Computer Science and Technology	Information Technology Laboratory
Institute for Material Science and Engineering	Materials Science and Engineering Laboratory
Institute for Materials Research	Materials Science and Engineering Laboratory
Laboratory for the Astrophysics Division	Physics Laboratory
Manufacturing Engineering Laboratory	Manufacturing Engineering Laboratory
Materials Science & Engineering Laboratory	Materials Science and Engineering Laboratory
National Engineering Laboratory	Multiple
National Institute of Standards and Technology	Multiple
National Measurement Laboratory	Multiple
Physics Laboratory	Physics Laboratory

Appendix H

THE NATIONAL ACADEMIES

Advisers to the Nation on Science, Engineering, and Medicine

National Research Council

Research Associateship Programs

FINAL REPORT

Return this form directly to the NRC as an E-mail attachment, or print out and mail or fax.

1) *Associate Last or Family Name*	**First Name**	*M.I.*
2) *FORWARDING Address* **(to which your tax statement will be mailed)** **Res. or Inst.** **Street** **City, State Zip**	*FORWARDING Phone(s) and E-Mail* **(if known)** **Home Phone:** **Alt. Phone:** **E-mail:**	
3) *Today's Date*	*Dates of Tenure* from to	

4) *Agency*	*Laboratory or Center*	**Division / Directorate / Department**

5) *Name of Laboratory NRC Adviser* **(and USMA Mentor, if applicable)**

6) *TITLE OF RESEARCH PROPOSAL*

7) *SUMMARY OF RESEARCH DURING TENURE* **Itemize significant findings in concise form, utilizing key concepts/words.**

1)
2)
3)
4)
5)

(USMA Davies Fellow: please add summary of teaching, including classes taught.)

8) *RESEARCH IN PROGRESS* **Describe in no more than 100 words.**

9) *PUBLICATIONS AND PAPERS RESULTING FROM NRC ASSOCIATESHIP RESEARCH*

Provide complete citations: author(s), title, full name of journal, volume number, page number(s), and year of publication.

a) Publications in peer-reviewed journals

b) Books, book chapters, other publications

c) Manuscripts in preparation, manuscripts submitted

10) *PATENT OR COPYRIGHT APPLICATIONS RESULTING FROM NRC ASSOCIATESHIP RESEARCH*
Provide titles, inventors, and dates of applications.

11) *PRESENTATIONS AT SCIENTIFIC MEETINGS OR CONFERENCES*
Provide complete references: author(s), title, abstract/proceeding citation, meeting name and location.

International

Domestic

12) *SEMINARS OR LECTURES DELIVERED AT UNIVERSITIES AND/OR INSTITUTES* **Include dates, names and locations of seminars.**

13) *PROFESSIONAL AWARDS RECEIVED DURING TENURE*

14) *POST-TENURE POSITION TITLE*

15) *POST-TENURE 0RGANIZATION* **Provide name and address of organization.**

16) *POST-TENURE POSITION STATUS / CATEGORY* **Please indicate only one.**

☐ Remain at Host Agency as Permanent Employee
☐ Remain at Host Agency as Contract/Temporary Employee
Abbreviate Host Laboratory/Center ______
☐ Research Position at Another US Government Laboratory
☐ Administrative Position at US Government Laboratory
☐ Research Position at Foreign Government Laboratory
☐ Research/Teaching at US College/University
☐ Research/Teaching at Foreign College/University
☐ Research/Administration in Industry
☐ Research/Administration in Non-Profit Organization
☐ Postdoctoral Research
☐ Self Employed
☐ Other: **specify** ______

17) *APPRAISAL OF RESEARCH ASSOCIATESHIP PROGRAM*

On a scale of 1 – 10 (poor - excellent), please rate the following:

SHORT TERM VALUE

☐ Development of knowledge, skills, and research productivity

Comments

LONG TERM VALUE

☐ How the NRC Associateship award affected your career to date

Comments

LAB SUPPORT

☐ Quality of support from the Laboratory--equipment, funding, orientation, safety and health guidelines, etc.

Comments

ADVISER/MENTOR SUPPORT

☐ Quality of mentoring from the Laboratory NRC Adviser (USMA Mentor, if applicable)

Comments

LPR SUPPORT

☐ Quality of administrative support from the **L**aboratory (e.g., NIST) NRC **P**rogram **R**epresentative (LPR)

Comments

NRC SUPPORT

☐ Quality of administrative support (applications, inquiries, post-review, award-related, travel, etc.) from the NRC

Comments

18) *PLEASE PROVIDE ANY SUGGESTIONS FOR PROGRAM IMPROVEMENT.*

Appendix I

THE NATIONAL ACADEMIES

Advisers to the Nation on Science, Engineering, and Medicine

Research Associateship Programs

EVALUATION of ASSOCIATE by ADVISER

If evaluation is for renewal, attach it to Associate's Renewal Application, and forward to LPR for signature.

If evaluation is for end of tenure, sign it and send directly to the coordinator at the NRC.

Associate Last or Family Name				*First Name*		*M.I.*
Agency	*Laboratory or Center*	*Eval. for* **yr.**	*Tenure Requested* **__ Months (if applicable)**	*Original Starting Date*	*Renewal* **applicable)**	*Date* **(if**
Adviser Last Name **(USMA Mentor, if applicable)**				*Adviser First Name*		M.I.

Adviser Laboratory Address **Division / Branch / Department**	*Adviser Laboratory Phone*
Address **Building, Room, Mail Stop / Code, if applicable**	*Fax*
Address **Street**	*Adviser E-mail*
City State Zip	

1) *Briefly, how long and in what capacity have you known the Associate?*

2) *Briefly,* **(e.g., half-page maximum)** *comment on the progress, sufficient time/schedule, and principal accomplishments of the research (or teaching, if USMA Davies Fellow), acknowledging, but not listing, publications. If this evaluation recommends extended tenure, please list specific reasons.*

3) *The purpose of the Research Associateship Programs is to provide to postdoctoral and senior scientists and engineers of unusual ability and promise an opportunity to conduct research on problems largely of their own choice which may contribute to the general research effort of the host laboratory. To what extent is this purpose is being fulfilled?*
(Not applicable to USMA Mentor)

4) *According to the categories below, please rate the Associate in comparison to scientists and engineers (or teaching professionals, if USMA Davies Fellow) with comparable training and experience.*

	Below Average	Average	Above Average	Good	Exceptional
Knowledge of Field	☐	☐	☐	☐	☐
Research (or Teaching) Techniques	☐	☐	☐	☐	☐
Motivation/Initiative	☐	☐	☐	☐	☐
Independent Research	☐	☐	☐	☐	☐
Innovative Thinking	☐	☐	☐	☐	☐
Overall Scientific (or Teaching) Ability	☐	☐	☐	☐	☐

5) *Has the Associate been effective in relationships with others in scientific matters? You may wish to comment on such attributes as leadership, cooperation, assertiveness, and influence on colleagues and other branches of the organization.*

6) *If the Associate is a productive scientist, what, in your opinion, is the quality of the work? If the Associate is not a productive scientist, why, in your opinion, is this the case?*

7) *Add any other pertinent comments that will help assess the Associate's ability and potential for research (or teaching, if USMA Davies Fellow). Please comment on weaknesses as well as on strong points.*

8) *Would you like the Associate as a professional colleague?*

NO ☐ **YES** ☐ **NO COMMENT** ☐

9) *Do you recommend the Associate's tenure be renewed?*

NO ☐ **YES** ☐ **NOT APPLICABLE** ☐

10) *Please provide any suggestions for program improvement*

Adviser Signature..Date...

USMA Mentor signature *(if applicable)*.......................................Date...........................

Laboratory Program Representative (LPR) Signature.................................Date..............
ONLY if this evaluation is for renewal